Finite elements: an introduction for engineers

Finite elements

*AN INTRODUCTION FOR
ENGINEERS*

R. K. LIVESLEY

*University lecturer in Engineering
and Fellow of Churchill College, Cambridge*

CAMBRIDGE UNIVERSITY PRESS

Cambridge

London New York New Rochelle

Melbourne Sydney

CAMBRIDGE UNIVERSITY PRESS
Cambridge, New York, Melbourne, Madrid, Cape Town, Singapore, São Paulo, Delhi

Cambridge University Press
The Edinburgh Building, Cambridge CB2 8RU, UK

Published in the United States of America by Cambridge University Press, New York

www.cambridge.org
Information on this title: www.cambridge.org/9780521285971

First published 1983
Re-issued in this digitally printed version 2009

A catalogue record for this publication is available from the British Library

Library of Congress Catalogue Card Number: 82-22155

ISBN 978-0-521-24314-8 hardback
ISBN 978-0-521-28597-1 paperback

CONTENTS

PREFACE

Finite elements were first used in the 1950s in aircraft design, and the method is now taught in many universities and technical colleges as a numerical technique for stress analysis. However, in recent years the approach has also been used increasingly in other branches of engineering. This widening of the area of application has revealed the finite-element method for what it really is – a general procedure for obtaining approximate solutions to elliptic partial differential equations, and has made it an appropriate topic for inclusion in an undergraduate course on engineering mathematics.

This view of the method has governed the arrangement of the present text, which is based on a course of lectures available to third-year students in the Cambridge University Engineering Department. The lectures are intended to provide a link between the traditional mathematical topics taught in the first two years of a general engineering degree course and specialist lectures associated with particular areas of application.

The book is intended for engineers who require a sound understanding of the mathematical basis of the finite-element method in preparation for the study of more specialised texts. It emphasises the relationship of the method to other techniques of approximate numerical analysis and uses variational ideas to establish convergence criteria. Examples are drawn from a number of engineering disciplines in order to emphasise the versatility of the approach. A number of references for background reading on related topics are given on p. 190.

I have concentrated on those ideas which I believe to be fundamental to the method. Inevitably much has had to be left out. There is no mention of non-linear materials, hybrid elements or gross-deformation analysis. The only variational principle used is the one which appears in continuum mechanics as the principle of minimum potential energy. Reduced integration and the patch test are described, but the emphasis is on solutions based on conforming elements, for which strict bounding principles exist.

The first chapter reviews a number of general mathematical ideas and techniques which are used later in the book. The next six chapters present the basic theoretical ideas of the finite-element method. Each of the first seven chapters includes a collection of problems, many of which lead to important extensions of the ideas presented in the main part of the text. The solutions to these problems are set out in full at the end of each chapter. The final chapter discusses the programming of the method. Although no problems are provided for this chapter, any reader seeking a balanced view of finite-element analysis is strongly advised to spend some time using a finite-element program. Ideally such use should be set in the context of a real engineering design problem. Listings of such programs may be found in a number of textbooks.

I have tried to describe *how* the finite-element works. I have not discussed the more general problem of *when* and *why* it should be used. The availability of large-scale commercial finite-element programs gives an engineer the option of obtaining very detailed information about the behaviour of the objects he designs. However, the existence of this option increases, rather than reduces, the need for sound engineering judgement on the part of the program user. A stack of computer output can never be a substitute for understanding and common-sense.

I am grateful to many people who have influenced, directly or indirectly, the contents and arrangement of this book. The first draft was written after reading a set of lecture notes prepared by Dr C. Szalwinski of the Cambridge University Soil Mechanics group. Geoffrey Butlin, David Livesley and Bill Stronge have read individual chapters and made useful comments. The description of the PAFEC system in chapter 8 is quoted by permission of PAFEC Ltd. The cover illustration, Figs. 3.5 and 8.7 were produced by FEGS Ltd and Shape Data Ltd. Finally, the production of the manuscript and a number of the drawings was done using text-editing and graphics facilities provided by the Cambridge University Engineering Department's computing service.

Cambridge, May 1982

1

Preliminaries

This book is concerned with *field problems*, that is, problems whose solution involves the solution of partial differential equations with appropriate boundary conditions. Such problems occur in a number of important areas of engineering science, including stress analysis, fluid and thermal flow, diffusion and electromagnetism.

The finite-element method is simply a numerical technique for obtaining an approximate solution to a field problem. It converts the governing differential equation into a set of linear algebraic equations, and its popularity rests largely on the ease with which these equations can be assembled and solved on a computer.

This chapter presents a number of ideas which are important for a sound understanding of the finite-element method. It begins with a brief review of some mathematical concepts and techniques which are used later in the book. It continues with accounts of the Rayleigh–Ritz and finite-difference methods, and concludes with a brief summary of the matrix approach to the analysis of electrical networks and skeletal structures.

1.1 Some mathematical building-bricks

Vector calculus is the natural language for discussing field problems. Not a great deal is required to follow this book, but the reader is assumed to be familiar with the concepts of *gradient* and *divergence*. The (vector) gradient of a scalar field u is denoted by ∇u and the (scalar) divergence of a vector field $\mathbf{v}$ by $\nabla \cdot \mathbf{v}$. Results such as $\nabla \cdot (u\mathbf{v}) = \nabla u \cdot \mathbf{v} + u\nabla \cdot \mathbf{v}$ are assumed, and occasional use is made of the divergence theorem. Most of the analysis is first developed for two-dimensional problems and then generalised to three-dimensional ones.

Matrix algebra is the natural language for the assembly of linear algebraic equations and is used extensively in all books on the finite-element method. Once again, not a great deal is assumed in the present text, the only matrix operations used being those of multiplication, addition and transposition. While most of the book is concerned with assembling equations rather than solving them, the reader is assumed to be familiar with the ordinary Gaussian elimination method of solving linear algebraic equations.

The use of vector calculus allows a field problem to be stated in a very compact way, without any need for the introduction of a specific coordinate system. In numerical solutions, however, vectors are normally expressed in component form. Operations involving such vectors are conveniently written using the notation of matrix algebra. Thus the scalar product of two vectors **a** and **b** may be written either as **a**·**b** or as $\mathbf{a}^t\mathbf{b}$, the convention in the second form being that $\mathbf{a}^t$ represents a *row* vector, while **b** represents a *column* vector, i.e.

$$\mathbf{a}^t\mathbf{b} = [a_x\, a_y\, a_z] \begin{bmatrix} b_x \\ b_y \\ b_z \end{bmatrix}$$

In the same way the gradient operator may be written as

$$\nabla = \begin{bmatrix} \partial/\partial x \\ \partial/\partial y \\ \partial/\partial z \end{bmatrix} \quad \text{(a column vector)}$$

the divergence operator being represented by ∇^t (a row vector).† The use of this notation emphasises the similarity of numerical methods developed for equations (such as those of Laplace and Poisson) involving the vector differential operator ∇ and their counterparts in linear stress analysis, where the corresponding differential operator is a matrix.

The finite-element method focusses attention on particular points or 'nodes' in the solution region, and much of the analysis refers to scalars, vectors or matrices associated with nodes. Such quantities have subscripts which indicate the node or nodes with which they are associated – in this book $i, j, ..., m$ will be used as general subscripts ranging over all the nodes of an element or a solution region, while p, q, ..., s will be used to pick out specific nodes. To save space, contracted notation will be used consistently: where a subscript is repeated within a multiplication, including multiplication by a differential operator, summation of the product is implied over the range of the subscript. A repeated subscript of this kind is referred to

† This simple transposition relationship does not hold in general curvilinear coordinates. See chapter 6 for a more detailed discussion of this point.

as a 'dummy subscript' and plays the same sort of role as the index variable in a Fortran 'DO' loop. Any analysis set out in contracted notation can always be checked by writing it out 'longhand'.

A recurrent theme of this book is the representation of a function by a polynomial approximation. The simplest form for such an approximation is

$$u(x) = c_0 + c_1 x + c_2 x^2 + \ldots + c_{M-1} x^{M-1} \qquad (1.1)$$

However, when written in this form the value of a coefficient (other than c_0) has no direct relationship to the value of $u(x)$ at a particular point. An alternative form of (1.1) which will be used extensively in this book is

$$u(x) = u_1 n_1(x) + u_2 n_2(x) + \ldots + u_M n_M(x) = u_i n_i(x) \qquad (1.2)$$

In this expression the *coefficients* u_i are the values of $u(x)$ at a set of M points x_i, while the *functions* $n_i(x)$ are polynomials of degree $M-1$ possessing the property $n_i(x_j) = \delta_{ij}$ – i.e. $n_i(x_j)$ is 1 if $i = j$ and 0 if $i \neq j$. The general form of the function $n_i(x)$ for M arbitrary points is

$$n_i(x) = \frac{(x-x_1)(x-x_2), \ldots, (x-x_M)}{(x_i-x_1)(x_i-x_2), \ldots, (x_i-x_M)} \quad \begin{matrix} \text{(omitting } (x-x_i)) \\ \text{(omitting } (x_i-x_i)) \end{matrix} \Bigg\} \quad (1.3)$$

Such functions are termed Lagrange interpolating polynomials.

Equation (1.2), in which a *function* $u(x)$ is represented as a linear combination of M independent *functions* $n_i(x)$, is analogous to the equation $\mathbf{u} = u_i \mathbf{n}_i$, in which a *vector* $\mathbf{u}$ is represented as a linear combination of M independent unit *vectors* $\mathbf{n}_i$, $\mathbf{u}$ and $\mathbf{n}_i$ being vectors in a Euclidean space of P dimensions $(P \geqslant M)$.

Polynomials of the type defined in (1.1) and (1.2) can be used in two different ways. The first is as easily-computed approximations to more complicated *known* functions. The form given in (1.2) is particularly useful in this respect. For if a given function $f(x)$ has values $f(x_j)$ at a set of M arbitrary points x_j, the approximation $u(x) = f(x_i) n_i(x)$ is equal to $f(x)$ whenever $x = x_j$, by virtue of the definition of the functions n_i.

As an example, consider the construction of a quadratic approximation $u(x)$ for a function $f(x)$ which has values $f(-1) = 5, f(0) = 8, f(1) = 13$. The functions n_i are

$$n_1 = (x-1)x/2, \quad n_2 = 1 - x^2, \quad n_3 = (x+1)x/2$$

These functions are shown in Fig. 1.1. It is easy to check that

$$u(x) = 5[(x-1)x/2] + 8[1-x^2] + 13[(x+1)x/2]$$

is a quadratic function (in fact *the* quadratic function) which takes the required values.

The second use of polynomial approximations, which is the one most relevant to the finite-element method, is as approximations to *unknown* functions. If a polynomial approximation $u(x) = u_i n_i(x)$ is substituted for the unknown solution of a differential equation, the problem is changed from that of finding an unknown continuous *function* to that of finding the finite set of *values u_i* which gives the best approximation to the solution.

The Lagrange interpolating polynomials $n_i(x)$ defined in (1.3) can easily be generalised to two and three dimensions. In this extended form they play a key role in the finite-element method, and are used extensively in later chapters of this book. There, as is customary in finite-element literature, they are referred to as *shape functions*.

The finite-element method often requires the evaluation of surface or volume integrals, the regions of integration being sub-divisions of the solution region – sub-divisions which are, in fact, the 'finite elements'. In such integrations the integrand is usually a matrix whose coefficients are functions of position. The integral of a matrix is simply a matrix of the same dimensions in which each coefficient is the integral of the corresponding coefficient in the integrand. This follows from the matrix addition rule and the fact that an integral can be regarded as the limit of a sum.

In the finite-element method the evaluation of these integrals is often carried out in two stages,

(*a*) A change of variable is made which maps the integration region into a geometrically simpler region – a triangle, square, tetrahedron, etc., as appropriate. A typical mapping for a curvilinear plane triangle is shown

Fig. 1.1. An approximation based on three quadratic polynomials.

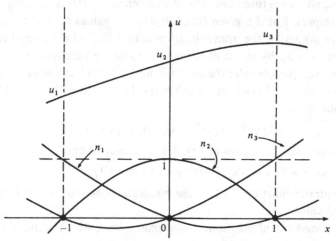

in Fig. 1.2. In the original x, y plane the variables α, β define a curvilinear coordinate system and are referred to as *parametric coordinates*. For a function $f(x, y)$ defined in a two-dimensional region such as the one shown in the figure the value of the integral of $f(x, y)$ is given by

$$\iint_R f(x, y)\, \mathrm{d}x\, \mathrm{d}y = \iint_{R'} f(x(\alpha, \beta), y(\alpha, \beta))|\mathbf{J}|\mathrm{d}\alpha\, \mathrm{d}\beta$$

where $|\mathbf{J}|$ is the determinant of the Jacobian matrix

$$\mathbf{J} = \begin{bmatrix} \partial x/\partial \alpha & \partial y/\partial \alpha \\ \partial x/\partial \beta & \partial y/\partial \beta \end{bmatrix}$$

This determinant may be thought of as the local area-magnification factor of the mapping. A similar formula applies to volume integrals.

(*b*) The transformed integral is evaluated numerically using, almost invariably, a Gauss integration formula. This is a two- or three-dimensional extension of the one-dimensional Gauss formula, which has the general form

$$\int_{-1}^{1} f(\alpha)\, \mathrm{d}\alpha \approx h_i\, f(\alpha_i) \quad (i = 1,\dots, N) \tag{1.4}$$

where the values α_i (the 'Gauss points') are the roots of the Legendre polynomial $P_N(\alpha)$ and the coefficients h_i (the 'Gauss weights') are constants depending only on the number N. Gauss integration formulae are more accurate than 'equal interval' formulae such as Simpson's rule, a one-dimensional N-point formula being exact for any polynomial of degree $2N-1$. For example, the one-dimensional 'three-point formula'

$$\int_{-1}^{1} f(\alpha)\, \mathrm{d}\alpha \approx [5f(-0.7746) + 8f(0) + 5f(0.7746)]/9$$

Fig. 1.2. A two-dimensional mapping: x and y are specified functions of α and β.

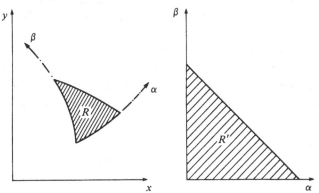

is exact if $f(\alpha)$ is any polynomial in α of degree 5 or less, while the two-dimensional 'four-point formula'

$$\int_{-1}^{1}\int_{-1}^{1} f(\alpha, \beta)\, d\alpha\, d\beta \approx f(0.5773, 0.5773) + f(0.5773, -0.5773)$$
$$+ f(-0.5773, 0.5773) + f(-0.5773, -0.5773)$$

is similarly exact for any cubic function.† Tables of Gauss points and weights for one-, two- and three-dimensional numerical integration will be found in reference 1, chapter 8.

1.2   Approximate solutions of differential equations – the Ritz method‡

Consider an ordinary linear differential equation $\mathscr{L}u(x) = w(x)$, where $\mathscr{L}$ is a linear differential operator (such as, for example, $x\, d^2/dx^2 + d/dx$), w is a known function of the independent variable x and sufficient boundary conditions are prescribed for a unique solution to exist in a range $a \leqslant x \leqslant b$. Imagine that no exact analytical expression for the solution can be found, and that an approximate solution $u(x)$ is to be constructed of the form

$$u(x) = c_i \phi_i(x) \quad (i = 1, ..., M) \tag{1.5}$$

where $\phi_1(x), ..., \phi_M(x)$ are a set of M linearly independent functions and $c_1, ..., c_M$ are constants which are to be determined.

The first step in the solution involves choosing the set of functions $\phi_i(x)$. One well-established approach is to choose a subset (usually the first M terms) of an infinite set of orthogonal functions – an M-term Fourier series is a familiar example. If the orthogonal functions are the eigenfunctions of the differential operator $\mathscr{L}$ (i.e. solutions of $\mathscr{L}\phi = \lambda\phi$), then a particularly simple form of solution can be obtained. However, it is often difficult to find a suitable set of orthogonal functions and the finite-element method (at least as described in this book) is based on non-orthogonal approximating functions which are simple polynomials – usually cubics at most.

Once the approximating functions have been chosen, the next step is to find the coefficients c_i which give the best approximation for that set of functions, and this requires some criterion for deciding what is meant by 'best'. The Ritz method provides one such criterion. It operates, not on

---

† A general cubic in two variables α and β has terms in 1, α, β, α^2, $\alpha\beta$, β^2, α^3, $\alpha^2\beta$, $\alpha\beta^2$, β^3.

‡ Historically this method is an extension, due to Ritz, of a method first proposed by Rayleigh.

the differential equation, but on the equivalent 'variational principle', in which the solution is associated with the stationary value of an integral. For differential equations derived from physical systems this integral often represents some form of energy, so that if the solution is a stable one the stationary value is a minimum.†

A problem which illustrates the method is that of finding the deflected shape of the suspended cable shown in Fig. 1.3 under a specified transverse loading $w(x)$. (It is assumed that the slope of the cable is everywhere small.) The true deflected shape $\tilde{u}(x)$ can be thought of as either,

(a) the solution of the differential equation $Hu'' = -w$, subject to the boundary conditions $u(0) = u(1) = 0$,

or (b) the function which, of all the functions $U(x)$ satisfying the boundary conditions given above, minimises the total potential energy,

$$T(U) = \int_0^1 H(U')^2/2 - wU \, dx \qquad (1.6)$$

The quantity $T(U)$ has a value which depends on the choice of function and is called a 'functional'. Note that the boundary conditions appear as restrictions on the class of functions which may be considered in the minimisation.

These two formulations of the problem are mathematically equivalent.

For this particular example the steps of the Ritz method are as follows.

(1) A set of M functions $\phi_i(x)$ is chosen such that each individual function ϕ_i satisfies the boundary conditions. This implies that any approximate solution of the form $u(x) = c_i\phi_i(x)$ also satisfies the boundary conditions and is therefore an 'admissible function' as far as formulation (b) is concerned. Substitution of this function into (1.6) gives

$$T(u) = \int_0^1 H(u')^2/2 - wu \, dx \qquad (1.7)$$

(2) The functional $T(u)$ is minimised with respect to each one of the

† A stable equilibrium state is always one of minimum energy.

Fig. 1.3. A flexible cable carrying an arbitrary transverse load $w(x)$.

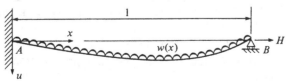

coefficients c_i, giving a set of M equations

$$\frac{\partial T}{\partial c_i} = \int_0^1 Hu' \frac{\partial u'}{\partial c_i} - w \frac{\partial u}{\partial c_i} \, dx = 0 \tag{1.8}$$

where i indicates any one of the subscripts $1, ..., M$, or

$$\int_0^1 H(c_j \phi_j') \frac{\partial(c_k \phi_k')}{\partial c_i} - w \frac{\partial(c_k \phi_k)}{\partial c_i} \, dx = 0 \tag{1.9}$$

Note the use of the independent pairs of dummy subscripts j and k to denote the two independent summations. Since the coefficients c_i are independent, $\partial c_k / \partial c_i$ equals 1 if $k = i$ and 0 if $k \neq i$. Thus (1.9) becomes

$$\left[H \int_0^1 \phi_i' \phi_j' \, dx \right] c_j = \int_0^1 w \phi_i \, dx \tag{1.10}$$

or

$$k_{ij} c_j = w_i \tag{1.11}$$

where

$$k_{ij} = H \int_0^1 \phi_i' \phi_j' \, dx \quad \text{and} \quad w_i = \int_0^1 w \phi_i \, dx$$

Equation (1.11) represents a set of M linear algebraic equations. From the form of (1.10) it follows that the equations are symmetric.

(3) Equations (1.11) are solved for the coefficients c_j.

As an example, consider the case $w(x) = Wx$, where W is a constant. Direct integration of the differential equation and insertion of the boundary conditions gives the exact solution $\tilde{u} = Wx(1-x^2)/6H$. Applying the Ritz process with $\phi_1 = x(1-x)$ and $\phi_2 = x^2(1-x)$ gives

$$k_{11} = H \int_0^1 (\phi_1')^2 \, dx = H/3, \quad k_{12} = k_{21} = H \int_0^1 \phi_1' \phi_2' \, dx = H/6$$

$$k_{22} = H \int_0^1 (\phi_2')^2 \, dx = 2H/15, \quad w_1 = W \int_0^1 \phi_1 x \, dx = W/12$$

$$w_2 = W \int_0^1 \phi_2 x \, dx = W/20$$

so that (1.11) becomes

$$\left. \begin{array}{l} c_1/3 + c_2/6 = W/12H \\ c_1/6 + 2c_2/15 = W/20H \end{array} \right\} \quad \text{with solution} \quad c_1 = c_2 = W/6H$$

giving $u = Wx(1-x^2)/6H$. The attainment of the exact solution in this particular case is due to the fact that the chosen approximation $u = c_i \phi_i$ includes that solution.

Equation (1.8) states that the best approximation is one which gives $T(u)$ a stationary value with respect to the coefficients c_i. An alternative way of stating this condition is to say that for a change of displacement Δu the change in $T(u)$ must be of second order. For the function $T(u)$ given in (1.7) this implies that

$$T(u+\Delta u) - T(u) = \int_0^1 [H(u' + \Delta u')^2 - H(u')^2]/2 - w\,\Delta u\,dx$$

must be of second order, so that

$$\int_0^1 Hu'\,\Delta u' - w\,\Delta u\,dx = 0 \qquad (1.12)$$

Equation (1.12) is equivalent to (1.8) and is recognisable as an equation of virtual work. The change Δu in the displaced form of the cable means, in general, that the point B in Fig. 1.3 moves horizontally, and the first term of (1.12) is the work done by the horizontal force H during that movement. (The proof of this statement is left as an exercise for the reader.) The second term of (1.12) is (rather more obviously) the work done by the transverse loading as the displacement of the cable changes. Note that the displacement Δu is not completely arbitrary, since it is produced by varying the c_is in equation (1.5). It therefore belongs, just as u itself belongs, to the class of functions defined by (1.5).

Virtual work equations are often used in the development of the finite-element method for stress analysis. It is important to realise that such equations are simply an alternative version of the energy minimisation equations used in the classical form of the Ritz method.

1.3   Conditions for convergence of the Ritz method

As the number of approximating functions ϕ_i increases, the solution given by the Ritz method converges to the true solution of a differential equation, provided that the approximating functions satisfy the following *sufficient* conditions.†

    (1) The infinite set of functions must be capable of representing the true solution exactly.

    (2) Each individual function must only give rise to finite terms in the functional to be minimised.

    (3) Each individual function must represent a physically acceptable solution – i.e. it must not violate the material continuity conditions implicit in the derivation of the differential equation.

    (4) Each individual function must satisfy the 'essential boundary

† Not all these conditions are always *necessary*. Some less restrictive conditions are discussed in section 5.6.

conditions' of the problem being solved,† though it need not satisfy the 'natural boundary conditions'. (These terms are explained later in this section.)

The implications of these restrictions are most easily understood by returning to the suspended-cable problem considered in the previous section.

The first condition is referred to in mathematical textbooks as the condition of *completeness*. It does not usually pose practical problems, though it is always necessary to make sure that the functions chosen have sufficient generality for the problem being solved. For example, if the functions ϕ_i are all symmetric about a certain point then the approximation u will inevitably be symmetric too, however many terms of the series (1.5) are taken.

The significance of restrictions 2 and 3 may be seen by considering the function ϕ shown in Fig. 1.4a. This is an example of a 'piecewise-linear' function. The problem is to decide whether the discontinuity of slope rules out the use of ϕ in a Ritz solution of the suspended cable problem of Fig. 1.3.

The effect of the discontinuity may be seen by starting with the 'smoothed' function ϕ_s shown in Fig. 1.4b and letting the 'transition interval' ε tend to zero. The smoothed function ϕ_s certainly satisfies restrictions 2 and 3, since it has continuous first derivatives, and the contribution

$$\int_0^1 H(\phi_s')^2/2 - w\phi_s \, dx \tag{1.13}$$

which it makes to the functional $T(u)$ is therefore finite. Furthermore, since ϕ_s' is finite whatever the value of ε, the contribution of the transition interval to (1.13) is of order ε and vanishes as $\varepsilon \to 0$. Hence the function ϕ gives rise to only finite terms in the functional and satisfies restrictions 2 and 3. It is, therefore, an admissible function. An alternative approach which leads to the same conclusion is based on the fact that if ϕ is to be an admissible part of some general deformation pattern $u = c_i\phi_i$ then there must be some *finite* loading which will produce *only* the deformation ϕ. In the case of Fig. 1.4a this loading is, of course, a single concentrated load.

In contrast, consider the use of the same functions in a Ritz analysis of a simply supported beam of unit length, uniform stiffness EI and carrying a transverse loading $w(x)$. The total potential energy associated with a

---

† Strictly this condition only applies in problems where each essential boundary condition is *homogeneous*, i.e. has zero on the right-hand side, as in $u(0) = 0$. The more general non-homogeneous case is discussed at the end of this section.

given displaced form $u(x)$ is

$$T(u) = \int_0^1 EI(u'')^2/2 - wu\,dx \tag{1.14}$$

where the first term of the integrand is the strain energy of bending and the second is the potential energy of the loading. The contribution which ϕ_s makes to $T(u)$ is

$$\int_0^1 EI(\phi_s'')^2/2 - w\phi_s\,dx \tag{1.15}$$

The smoothed function ϕ_s has finite second derivative and satisfies conditions 2 and 3. However, the curvature of the beam within the transition region is c/ε, which makes the contribution of the transition region to the first term of the integral in (1.15) equal to $EIc^2/2\varepsilon$. Consequently the contribution (1.15) tends to infinity as $\varepsilon \to 0$. Thus the function ϕ of Fig. 1.4*a* does not satisfy condition 2 and is not an admissible function in this case. This conclusion may also be reached by noting that

Fig. 1.4. (*a*) A piecewise-linear function with a discontinuity of slope. (*b*) The discontinuity removed by the insertion of a transition interval ε.

(*a*)

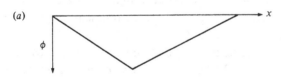

(*b*)

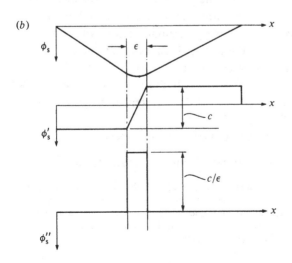

the external loading required to deform the beam to the shape of Fig. 1.4*a* is a pair of equal and opposite moments of infinite magnitude, applied to the two sides of the discontinuity in slope.

It is tempting to substitute the piecewise-linear function ϕ directly into (1.14) and ignore the contribution of the discontinuity in ϕ'' to the strain energy of bending. However, this is physically equivalent to allowing a frictionless hinge to occur in the beam at the point of discontinuity. Although this makes the associated strain energy finite, the function ϕ now violates condition 3 and is still not admissible. [As one might expect, its use in a Ritz analysis will indicate that a beam with a hinge at a point along its length cannot carry any load.]

From this example may be deduced the general rule that, *to ensure convergence, functions used in the Ritz method should have continuity of value and derivatives up to an order one less than the highest order of derivative appearing in the expression to be minimised.* This result also holds for systems with more than one independent variable. Problems in which the functional only involves *first* derivatives require approximating functions which are continuous in *value*. Such problems arise in potential theory (see chapter 2) and elastic stress analysis† (see chapter 3). Problems in which the functional involves *second* derivatives require approximating functions which are continuous in *value* and *gradient*. Such problems arise in the theory of plates and shells in bending (see chapter 7).

Condition 4 introduces the terms 'essential boundary conditions' and 'natural boundary conditions'. Essential boundary conditions (provided they are homogeneous) must be satisfied by each individual approximating function ϕ if convergence to the true solution is to be ensured. Natural boundary conditions give rise to terms in the functional and are therefore satisfied, at least approximately, during the minimisation, with convergence to the correct values in the limit. The difference may be illustrated by changing the problem discussed in section 1.2 to that shown in Fig. 1.5, where F, H and $w(x)$ are assumed known. In this problem the displacement boundary condition $u(1) = 0$ of Fig. 1.3 is replaced by a condition specifying the value of $u'(1)$.

The two alternative formulations of section 1.2 become,

(*a*) solve the differential equation $Hu'' = -w$ subject to the (homo-

---

† Provided that the functional is constructed with *displacement* as the dependent variable. This is the usual procedure in the application of the finite-element method to continuum stress analysis. If the functional is written in terms of a *stress function* then second derivatives appear and the analysis becomes very similar to that associated with the bending of thin plates.

geneous) essential boundary condition $u(0) = 0$ and the natural boundary condition $u'(1) = F/H$,

or (*b*) find the function which minimises the total potential energy

$$T(U) = \int_0^1 H(U')^2/2 - wU \, dx - FU(1) \tag{1.16}$$

considering only those functions $U(x)$ which satisfy the essential boundary condition $U(0) = 0$.

The central point here is that the term $FU(1)$ in (1.16) provides information about the natural boundary condition $u'(1) = F/H$ *within* the energy expression, whereas there is nothing in (1.16) corresponding to the essential boundary condition $u(0) = 0$. The latter condition must therefore be inserted *explicitly* when the form of the approximating function is chosen. For the particular case $w = Wx$ considered previously it is easy to verify that the Ritz process gives the exact solution $\tilde{u}(x) = x[(F + W/2) - Wx^2/6]/H$ from an approximating function $u(x) = c_1 x + c_2 x^2 + c_3 x^3$ – i.e. a function whose individual terms satisfy the essential boundary condition but not the natural boundary condition.

A similar distinction may be made in the application of the Ritz process to the bending of beams, plates and shells. In such problems essential boundary conditions are conditions which specify displacement or slope, while natural boundary conditions are those which specify shear or moment. In general the essential boundary conditions are conditions on the function and its derivatives up to an order one less than the highest order of derivative appearing in the expression to be minimised.

The Ritz procedure described in section 1.2 must be modified slightly if the essential boundary conditions are not homogeneous. In such cases the approximation (1.5) must be written in the form

$$u(x) = \phi_0(x) + c_i \phi_i(x) \quad (i = 1, \ldots, M) \tag{1.17}$$

Fig. 1.5. An example showing essential and natural boundary conditions.

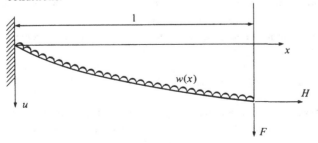

where $\phi_0(x)$ satisfies the essential boundary conditions, as specified, while the functions $\phi_i(x)$ satisfy them *in their homogeneous form*. For example, if the suspended-cable problem discussed in section 1.2 is changed so that the boundary conditions are $u(0) = 0$, $u(1) = C$, then a suitable approximating function is

$$u(x) = Cx + c_i\phi_i(x)$$

where the functions $\phi_i(x)$ satisfy the homogeneous boundary conditions $u(0) = u(1) = 0$. It is clear that this form of the function $u(x)$ satisfies the correct boundary conditions for all values of the coefficients c_i.

The inclusion in the set $U(x)$ of functions which do not satisfy the essential boundary conditions is, in effect, a relaxation of the constraints on the original physical system. For example, if an unknown constant displacement term is included in the displacement u associated with Fig. 1.5, this allows the whole system to undergo arbitrary rigid-body displacement. In the same way, if a cantilever with its clamped end at the origin is permitted a displacement under transverse load of the form $u = cx$, where c is a variable parameter, this is equivalent to the insertion of a hinge at the support and the Ritz process cannot be expected to give a meaningful result.

1.4   Bounds satisfied by Ritz solutions

As mentioned in section 1.2, a stable equilibrium state is always one of minimum energy. This means that the total potential energy $T(u)$ associated with a Ritz approximation u is always greater than the total potential energy $T(\tilde{u})$ associated with the true solution $\tilde{u}$, provided that the latter solution represents a stable state. If the functions ϕ_i are the first M terms of an infinite set satisfying the conditions given in the previous section then the approximation $T(u)$ will decrease monotonically to $T(\tilde{u})$ as $M \to \infty$. In other words, $T(u)$ is an *upper bound* on the true value of the total potential energy.

The inequality $T(u) \geqslant T(\tilde{u})$ may be converted into a more direct relationship between u and $\tilde{u}$. In the case of the suspended-cable example of section 1.2 the analysis is as follows.

The total potential energy associated with the true displaced form $\tilde{u}(x)$ is

$$T(\tilde{u}) = \int_0^1 H(\tilde{u}')^2/2 - w\tilde{u} \, dx \tag{1.18}$$

If the transverse loading is increased proportionally from zero to $w(x)$ then the work done by this loading must equal the work done by the cable against the constant horizontal force H. Thus

$$\int_0^1 w\tilde{u}/2 \, dx = H \int_0^1 (\tilde{u}')^2/2 \, dx \qquad (1.19)$$

and combining (1.18) and (1.19) gives

$$T(\tilde{u}) = -\int_0^1 w\tilde{u}/2 \, dx \qquad (1.20)$$

The total potential energy associated with the approximate displaced form $u(x)$ is

$$T(u) = \int_0^1 H(u')^2/2 - wu \, dx \qquad (1.21)$$

Since u is a Ritz approximation it satisfies the virtual work equation (1.12). Setting $\Delta u = u/2$ in that equation gives

$$H \int_0^1 (u')^2/2 \, dx = \int_0^1 wu/2 \, dx \qquad (1.22)$$

which when combined with (1.21) gives

$$T(u) = -\int_0^1 wu/2 \, dx \qquad (1.23)$$

Since $T(u) \geqslant T(\tilde{u})$ it follows from (1.20) and (1.23) that

$$\int_0^1 wu/2 \, dx \leqslant \int_0^1 w\tilde{u}/2 \, dx \qquad (1.24)$$

[Note that equation (1.22) cannot be derived by the direct 'work' argument used to obtain (1.19), since u is not the true equilibrium configuration. Indeed, equation (1.23) is not true unless u is a Ritz approximation.]

The inequality (1.24) states that the applied loading does more work in the true displacement than in any approximate displacement calculated by the Ritz process. In the case of a cable carrying a single concentrated load W at $x = a$ the inequality reduces to

$$u(a) \leqslant \tilde{u}(a) \qquad (1.25)$$

A similar analysis may be developed for any differential equation which can be solved approximately by the Ritz method. Equation (1.25) is often generalised to the statement that in any linear elastic system acted on by a single concentrated load, a Ritz approximation gives a *lower bound* to the true displacement under the load. However, this generalisation is somewhat less useful than it appears, since (in contrast to the one-dimensional example discussed in section 1.2), the application of a concentrated load to a two- or three-dimensional elastic continuum produces an infinite displacement at the loading point. It is better to regard a 'concentrated' load as being really a distributed load acting over a small but finite part of the solution region. Equation (1.24) or its generalisation may then be thought of as a statement about the *average* displacement of the loading zone.

1.5   Other criteria for determining the coefficients c_i

The Ritz procedure, as described in section 1.2, requires a field problem to be set up as an integral minimisation. Thus it can only be applied to problems for which a variational principle exists. There are, however, other methods of determining the unknown coefficients in the approximating function which operate directly on the governing differential equation.

Consider, as before, a linear differential equation $\mathscr{L}u(x) = w(x)$, with sufficient boundary conditions prescribed to ensure a unique solution in the range $a \leqslant x \leqslant b$. Let $u = c_i \phi_i$ be an approximate solution, with M unknown coefficients c_i, in which each of the functions ϕ_i satisfies both the essential *and* the natural boundary conditions. The 'best' solution will be one in which the 'error function' $\varepsilon(x) = \mathscr{L}u - w$ is as small as possible, where once again it is a matter of defining a suitable measure of smallness.

Three alternative ways of obtaining a set of M linear algebraic equations for the c_i are,

(a) Make $\varepsilon(x)$ zero at M arbitrarily chosen points x_i within the solution range. This method is known as Collocation.

(b) Minimise the mean square error, i.e. minimise $\int_a^b [\varepsilon(x)]^2 \, dx$ with respect to the coefficients c_i.

(c) Make $\varepsilon(x)$ satisfy M conditions of the form

$$\int_a^b \varepsilon(x) \, \phi_i(x) \, dx = 0 \quad (i = 1, ..., M)$$

Substituting for $\varepsilon(x)$ gives

$$\int_a^b (c_j \mathscr{L}\phi_j - w) \, \phi_i \, dx = 0$$

or

$$\left[\int_a^b (\mathscr{L}\phi_j) \, \phi_i \, dx \right] c_j = \int_a^b w \phi_i \, dx \qquad (1.26)$$

This is known as Galerkin's method.

These three procedures are all special cases of a more general procedure in which the error function is made to satisfy M conditions of the form

$$\int_a^b \varepsilon(x)\,\gamma_i(x)\,dx = 0 \quad (i = 1, ..., M) \tag{1.27}$$

where the γ_i are a set of M 'weighting functions'. Choosing delta functions† $\delta(x_i, x)$ for the γ_i gives case (a), choosing $\gamma_i = \mathscr{L}\phi_i$ gives case (b), while choosing $\gamma_i = \phi_i$ gives case (c).

Although they appear different, the equations (1.26) associated with Galerkin's method are really just an alternative form of the Ritz equations. In the case of the suspended-cable example considered in section 1.2 this equivalence is easily demonstrated. The governing differential equation is $Hu'' = -w$, so that equation (1.26) reduces to

$$\left[-H\int_0^1 \phi_j'' \phi_i \,dx \right] c_j = \int_0^1 w\phi_i \,dx \tag{1.28}$$

Integration of the left-hand side by parts gives

$$\left[H\int_0^1 \phi_j' \phi_i' \,dx - H\phi_j' \phi_i \Big|_0^1 \right] c_j = \int w\phi_i \,dx$$

and the second of the terms in brackets is zero since ϕ_i is required to satisfy the essential boundary conditions $u(0) = u(1) = 0$. Hence (1.28) may be written as

$$\left[H\int_0^1 \phi_i' \phi_j' \,dx \right] c_j = \int_0^1 w\phi_i \,dx$$

which is identical to the equations (1.10) set up by the Ritz method.

This identity stems from the fact that in any problem in which the basic differential equation is a statement about local equilibrium, both the Ritz and Galerkin equations are essentially equations of virtual work. The equivalence of the Ritz equations (1.8) and the virtual work equations (1.12) has already been discussed. That the Galerkin equations (1.26) are also virtual work equations follows from the fact that the error function ε is a measure of the extent to which the approximation u fails to satisfy equilibrium. Thus equation (1.27) may be thought of as a statement that the residual force system $\varepsilon(x)$ does no work in any of the displacements defined by the functions γ_i, which in the Galerkin method are simply the functions ϕ_i used to build up u.

Since the Galerkin and Ritz processes are simply alternative ways of setting up the same equations the arguments concerning convergence and

---

† The delta-function $\delta(x_i, x)$ is zero for all values of x except within an infinitesimal interval enclosing $x = x_i$, where it takes a value such that $\int_{-\infty}^{\infty} \delta(x_i, x)\,dx = 1$.

continuity in section 1.3 apply equally to both, except that in the Galerkin method the natural boundary conditions must be inserted explicitly. The smoothing procedure of section 1.3 may be used in cases where the functions ϕ_i have discontinuities of derivative to check the validity of steps such as the integration by parts of equation (1.28).

1.6   Approximate solutions of differential equations – the method of finite differences

The various approximate procedures described in the previous sections are all based on a continuous approximating function defined at all points within the solution region of the differential equation. In contrast, the *finite-difference* method calculates approximate values of the solution at a finite set of points, other values being filled in afterwards by interpolation. The essence of the method is the replacement of the differential operator in the differential equation by a *difference* operator.

As an illustration of the method, consider again the suspended-cable problem used to demonstrate the Ritz procedure in section 1.2. The 'solution' which the finite-difference method seeks to determine is the set of displacements u_i associated with a set of points with coordinates x_i along the length of the cable. It is convenient to space the points equally, as shown in Fig. 1.6. Let there be $M+1$ points $x_0, ..., x_M$, where $x_0 = 0, x_M = 1$, the distance between successive points being $h = 1/M$.

Consider now three successive points x_{i-1}, x_i, x_{i+1}, at which the solution has values u_{i-1}, u_i, u_{i+1}. The first derivative u' at $x = x_i+h/2$ is approximately $(u_{i+1}-u_i)/h$, while the first derivative at $x = x_i-h/2$ is similarly approximately $(u_i-u_{i-1})/h$. Repeating the process gives an approximation for the second derivative at $x = x_i$ as

$$u'' \approx (u_{i-1}-2u_i+u_{i+1})/h^2 \qquad (1.29)$$

Fig. 1.6. Notation for solution of the flexible cable problem by the method of finite differences.

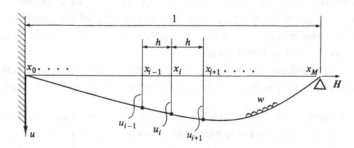

Substitution of (1.29) in the differential equation $Hu'' = -w$ gives a set of difference equations

$$H(-u_{i-1}+2u_i-u_{i+1})/h^2 = w(x_i) \quad (i = 1, ..., M-1) \tag{1.30}$$

where the boundary conditions on the differential equation imply $u_0 = u_M = 0$. Equations (1.30) may be written in matrix form as

$$(H/h^2) \begin{bmatrix} 2 & -1 & & & \\ -1 & 2 & -1 & & \\ & & \cdot & \cdot & \cdot \\ & & & \cdot & \cdot & \cdot \\ & & & & -1 & 2 \end{bmatrix} \begin{bmatrix} u_1 \\ \cdot \\ \cdot \\ \cdot \\ u_{M-1} \end{bmatrix} = \begin{bmatrix} w_1 \\ \cdot \\ \cdot \\ \cdot \\ w_{M-1} \end{bmatrix} \tag{1.31}$$

where $w_i = w(x_i)$. Note that each equation only involves three neighbouring values of u_i. These equations can now be solved for the unknowns $u_1, ..., u_{M-1}$. Note that this 'solution' gives no direct information about the displacements at other points on the cable.

The difference approximation (1.29) is often represented by a 'molecule' of coefficients, as shown in Fig. 1.7a. Similar difference formulae exist for partial differential operators. If a square lattice is defined in a two-dimensional region then an approximation to the operator ∇^2 is given by the molecule shown in Fig. 1.7b, while in a lattice of equilateral triangles the same operator is represented by the molecule in Fig. 1.7c. More

Fig. 1.7. Finite difference molecules:
(a) for d^2/dx^2.
(b) and (c) for ∇^2 in a plane.

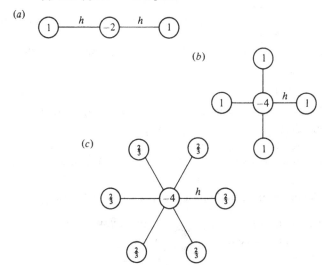

complex molecules can be developed for irregular meshes and higher-order operators.

1.7   Electrical and structural network analysis

When computers first became available in the early 1950's one of the earliest applications was to the analysis of electrical networks and structural frameworks. Procedures were developed for the systematic assembly of the linear equations describing these problems, and it is these procedures which are used in assembling the linear equations associated with the finite-element method.

Consider first a linear D.C. network. Let there be $M+1$ junctions or *nodes*, labelled $0, 1, ..., M$, node 0 being earthed. The nodes are connected by *elements*, each element being electrically equivalent to a resistor in parallel with a current source of known magnitude, as shown in Fig. 1.8a. The element connecting nodes p and q is shown isolated in Fig. 1.8b. Note that this element has *local* node numbers 1, 2 which correspond to *global* node numbers p, q in the assembled network. These two ways of numbering nodes will be encountered repeatedly in later chapters of this book.

It is convenient to consider the resistor and the current source separately, as shown in Fig. 1.8b. The relationship between the currents i_1, i_2 and the voltages v_1, v_2 is

$$i_1 = -i_2 = (v_1 - v_2)/r$$

which may be written as

$$\begin{bmatrix} i_1 \\ i_2 \end{bmatrix} = \begin{bmatrix} g_{11} & g_{12} \\ g_{21} & g_{22} \end{bmatrix} \begin{bmatrix} v_1 \\ v_2 \end{bmatrix} \qquad (1.32a)$$

where $g_{11} = g_{22} = -g_{12} = -g_{21} = 1/r$. These equations may be written in contracted notation as

$$i_i = g_{ij} v_j \quad (i, j = 1, 2) \qquad (1.32b)$$

Equations (1.32) are singular, since v_1 and v_2 may be changed by the same amount without altering the currents i_1 and i_2.

The current source may be regarded as being connected directly to nodes p and q, causing known currents $I_1 (= I)$ and $I_2 (= -I)$ to flow into those nodes.

Adding up the currents at node p of the assembled network gives

$$\sum \begin{matrix} \text{currents flowing into} \\ \text{resistors connected} \\ \text{to node p} \end{matrix} = \sum \begin{matrix} \text{currents flowing out} \\ \text{of sources connected} \\ \text{to node p} \end{matrix} \qquad (1.33)$$

Substitution from (1.32) into (1.33) for each resistor connected to node p gives a linear equation which must be satisfied by the nodal voltages at node p and its immediate neighbours. A similar procedure applied to each node gives a set of equations for the nodal voltages in the network. These equations may be constructed by carrying out the following steps:

(a) Set up an $M \times M$ matrix of zeros for the coefficients and a vector of M zeros for the right-hand side.

(b) For the element connecting global nodes p and q add the coefficients g_{ij} to the matrix and the currents I_1, I_2 to the vector in the positions shown in (1.34).

Fig. 1.8. (*a*) Part of a D.C. electrical network.
(*b*) A single element of a D.C. network.

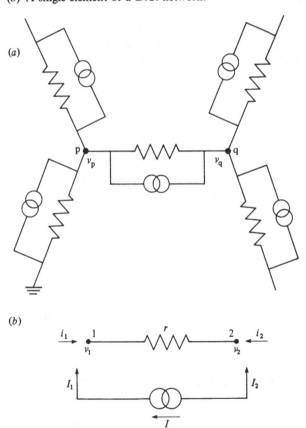

<table>
<tr><td></td><td colspan="2" align="center">Add to left-hand side
of equations</td><td align="center">Add to
right-hand
side of
equations</td><td></td></tr>
<tr><td></td><td align="center">column p</td><td align="center">column q</td><td></td><td></td></tr>
</table>

$$
\begin{matrix} \text{row p} \\ \\ \text{row q} \end{matrix}
\begin{bmatrix} \vdots & & \vdots & \\ \cdots & g_{11} & \cdots & g_{12} & \cdots \\ \vdots & & \vdots & \\ \cdots & g_{21} & \cdots & g_{22} & \cdots \\ \vdots & & \vdots & \end{bmatrix}
\begin{bmatrix} \vdots \\ v_p \\ \vdots \\ v_q \\ \vdots \end{bmatrix}
\begin{bmatrix} \vdots \\ I_1 \\ \vdots \\ I_2 \\ \vdots \end{bmatrix}
\qquad (1.34)
$$

(c) Repeat (b) for all elements. Elements which have one end connected to earth (node 0) only contribute one coefficient to the matrix and one term

Fig. 1.9. (a) Part of a pin-jointed framework.
(b) A single element of a pin-jointed framework.

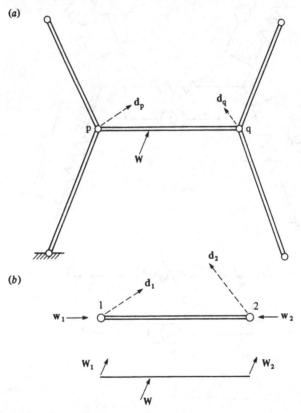

to the right-hand side, these being g_{11} and I_1 if end 2 is earthed and g_{22} and I_2 if end 1 is earthed.

Although equations (1.32) are singular the equations for the complete network are non-singular, provided that at least one node of the network is earthed. After solution of these equations for the unknown nodal voltages, the currents in the individual resistors may be obtained from equations (1.32).

The analysis of a pin-jointed truss follows exactly the same pattern, with voltages replaced by displacements and currents replaced by forces. (The only difference is that the displacements and forces are vectors, not scalars.) For the element shown in Fig. 1.9*b* the equations are

$$\begin{bmatrix} \mathbf{w}_1 \\ \mathbf{w}_2 \end{bmatrix} = \begin{bmatrix} \mathbf{K}_{11} & \mathbf{K}_{12} \\ \mathbf{K}_{21} & \mathbf{K}_{22} \end{bmatrix} \begin{bmatrix} \mathbf{d}_1 \\ \mathbf{d}_2 \end{bmatrix} \tag{1.35a}$$

or

$$\mathbf{w}_i = \mathbf{K}_{ij}\mathbf{d}_j \quad (i, j = 1, 2) \tag{1.35b}$$

where the four matrices $\mathbf{K}_{ij}$ are 2×2 matrices (in the case of plane trusses) or 3×3 matrices (in the case of space trusses) involving the axial stiffness of the bar and the direction cosines defining its orientation. Equations (1.35), like (1.32), are singular, since an isolated bar can be given a rigid-body displacement without affecting the end forces. The load $\mathbf{W}$ acting on the bar is replaced by equivalent loads $\mathbf{W}_1$, $\mathbf{W}_2$ acting at the nodes p, q. The assembly procedure is similar to that given for a D.C. network, with (1.34) being replaced by

	Add to left-hand side of equations			Add to right-hand side of equations
	column p	column q		
row p	$\cdots$ $\mathbf{K}_{11}$ $\cdots$	$\mathbf{K}_{12}$ $\cdots$	$\mathbf{d}_\mathrm{p}$	$\mathbf{W}_1$
row q	$\cdots$ $\mathbf{K}_{21}$ $\cdots$	$\mathbf{K}_{22}$ $\cdots$	$\mathbf{d}_\mathrm{q}$	$\mathbf{W}_2$

$$\tag{1.36}$$

As before, elements with end 1 connected to a fixed foundation only contribute $\mathbf{K}_{22}$ and $\mathbf{W}_2$ to (1.36). Similarly, elements with end 2 fixed only contribute $\mathbf{K}_{11}$ and $\mathbf{W}_1$. The nodal equations for the complete truss are non-singular, provided that the truss is adequately restrained against rigid-body displacement.

Problems for chapter 1

1.1 Write a computer program (in Fortran or a similar language) for the solution of the system of linear algebraic equations $\mathbf{Ax} = \mathbf{b}$, using the method of Gaussian elimination. [The equations should be treated sequentially and the leading diagonal element taken as pivot at each stage. No test for a zero pivot is required.]

1.2 Write down the cubic Lagrange interpolating polynomials associated with the four points $x_i = 0, a, 1-a, 1$. Hence or otherwise find the cubic interpolating polynomials ϕ_i associated with the polynomial approximation

$$u(x) = f(0)\,\phi_1 + f'(0)\,\phi_2 + f(1)\,\phi_3 + f'(1)\,\phi_4$$

which has the same value and derivative as the function $f(x)$ at $x = 0, 1$.

1.3 A region R has four boundaries

$$x(1-y) = 1, \quad x(1-y) = 2, \quad xy = 1, \quad xy = 3$$

Find variables α, β which map R into a rectangle in the α, β plane, and hence evaluate

$$\iint_R x^2 y \, dx \, dy$$

1.4 Find the ordinates α_i and the weights h_i which make the approximate integration formula

$$\int_{-1}^{1} f(\alpha) \, d\alpha \approx h_i f(\alpha_i) \quad (i = 1, \ldots, 3)$$

exact for any polynomial in α of degree 5 or less. [Since the formula is linear in f it is sufficient to make it exact for each of the functions $f(\alpha) = 1$, $\alpha, \alpha^2, \ldots, \alpha^5$.]

1.5 Use the Ritz method to find an approximate solution to the suspended-cable problem solved in section 1.2, using the one-term approximation $u = c_1 x(1-x)$.

Verify that $T(u) > T(\tilde{u})$, where $\tilde{u}$ is the true solution.

1.6 A uniform rod of unstretched length L and mass ρA per unit length hangs vertically from a fixed point. A mass ρAL is attached to the lower end. The vertical displacement of a point on the rod distant x from the fixed end is $u(x)$.

What are the essential and natural boundary conditions on u? Use the Ritz method to find an approximation for u using approximating functions of your own choice. Compare your approximate solution with the true solution $\tilde{u} = (\rho g L/E)\, x(2 - x/2L)$.

1.7 A uniform cantilever of unit length is built-in at $x = 0$ and carries a uniform transverse load of such magnitude that the governing differential equation for the transverse displacement u is

$$d^4u/dx^4 = 1$$

An approximate solution u takes the form $u = c_i \phi_i(x)$, where the functions ϕ_i satisfy the essential boundary conditions $\phi_i(0) = \phi'_i(0) = 0$. Show that the Ritz equations are

$$\int_0^1 \phi''_i \phi''_j c_j - \phi_i \, dx = 0$$

Find the coefficients c_j in the three approximate solutions

$$u^{(1)} = c_2 x^2$$
$$u^{(2)} = c_2 x^2 + c_3 x^3$$
$$u^{(3)} = c_2 x^2 + c_3 x^3 + c_4 x^4$$

Show that $u^{(3)}$ is the exact solution, and that $T(u^{(1)}) > T(u^{(2)}) > T(u^{(3)})$.

1.8 A conductor of length L has uniform resistance r per unit length and leakage conductance to earth g per unit length. Show that the potential v at a point distant x from one end satisfies the differential equation

$$v'' - grv = 0$$

Write down an expression for the power dissipated in the conductor and its surroundings due to a distribution of potential $v(x)$ which has specified values v_0, v_L at the ends of the conductor but is otherwise arbitrary. Show that the solution of the differential equation with boundary conditions $v(0) = v_0, v(L) = v_L$ minimises the expression for the power dissipated.

1.9 Apply the matrix assembly process of section 1.7 to the network shown in Fig. 1.10, indicating zero elements of the final matrix by a 0 and non-zero elements by an X.

Some equation-solving routines are most efficient if the distance of the non-zero elements from the leading diagonal of the matrix is as small as possible. Re-arrange the node numbering given in the figure to minimise the maximum value of this distance.

Solutions to problems

1.1 The procedure known as Gaussian elimination has two phases. In its simplest form the first phase of a procedure for N equations is as follows. Suitable multiples of the first equation are added to equations $2, ..., N$ to reduce the coefficients of x_1 in those equations to zero. The second equation is then used in the same manner to reduce the coefficients of x_2 in equations $3, ..., N$ to zero, and so on. This phase eventually transforms the original set of N equations into a set in which equation i only involves the variables $x_i, ..., x_N$. In Fortran this result can be achieved by the sequence

```
    DIMENSION A( , ),B( ),X( )
    DO 2 I=1,N−1
    DO 2 J=I+1,N
    Z=A(J,I)/A(I,I)
    DO 1 K=I,N
1   A(J,K)=A(J,K)−Z*A(I,K)
2   B(J)=B(J)−Z*B(I)
```

The second phase, the 'back substitution', begins with the solution of the N'th equation, which only involves x_N. The value obtained for x_N is substituted into the $N-1$'th equation, which is solved for x_{N-1}, and so on. A suitable Fortran sequence is

```
    DO 4 I=N,1,−1
    REPEAT 3, FOR J=(I+1,N)
3   B(I)=B(I)−A(I,J)*X(J)
4   X(I)=B(I)/A(I,I)
```

(The REPEAT statement is used in place of the more usual DO statement to avoid the execution of the statement with label 3 when I = N.)

1.2 For the given set of points x_i equation (1.3) gives

$$n_1 = (x-a)(x-1+a)(1-x)/a(1-a)$$
$$n_2 = x(x-1+a)(1-x)/a(2a-1)(1-a)$$
$$n_3 = x(x-a)(1-x)/(1-a)(1-2a)a$$
$$n_4 = x(x-a)(x-1+a)/a(1-a)$$

Fig. 1.10.

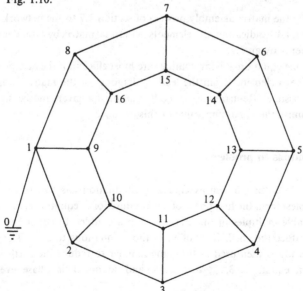

These expressions may be checked by substituting $x = 0, a, 1-a, 1$.

The expression $u = u(0) n_1 + u(a) n_2 + u(1-a) n_3 + u(1) n_4$ may be written as

$$u = \left(\frac{u(a)+u(0)}{2}\right)(n_2+n_1) + \left(\frac{u(a)-u(0)}{a}\right)\left(\frac{n_2-n_1}{2}\right)a$$
$$+ \left(\frac{u(1)+u(1-a)}{2}\right)(n_4+n_3) + \left(\frac{u(1)-u(1-a)}{a}\right)\left(\frac{n_4-n_3}{2}\right)a$$

Letting $a \to 0$ gives $u = u(0)\,\phi_1 + u'(0)\,\phi_2 + u(1)\,\phi_3 + u'(1)\,\phi_4$, where

$$\phi_1 = \operatorname*{Lim}_{a\to 0}\,(n_2+n_1) = 2x^3 - 3x^2 + 1$$

$$\phi_2 = \operatorname*{Lim}_{a\to 0}\left(\frac{n_2-n_1}{2}\right)a = x^3 - 2x^2 + x$$

$$\phi_3 = \operatorname*{Lim}_{a\to 0}\,(n_4+n_3) = 3x^2 - 2x^3$$

$$\phi_4 = \operatorname*{Lim}_{a\to 0}\left(\frac{n_4-n_3}{2}\right)a = x^3 - x^2$$

The functions $\phi_1, \ldots, \phi_4$ are known as *Hermite* polynomials and are shown in Fig. 1.11.

1.3 If new variables $\alpha = x(1-y), \beta = xy$ are defined, the regions of integration in the xy and $\alpha\beta$ planes are those shown in Fig. 1.12. The required integral may be evaluated from the relationship

$$\iint_R f(x,y)\,\mathrm{d}x\,\mathrm{d}y = \iint_{R'} f(x(\alpha,\beta), y(\alpha,\beta))|J|\,\mathrm{d}\alpha\,\mathrm{d}\beta$$

It is straightforward to express x and y in terms of α and β and compute $|J|$ directly. However, it is easier to form

$$J^{-1} = \begin{bmatrix} \partial\alpha/\partial x & \partial\alpha/\partial y \\ \partial\beta/\partial x & \partial\beta/\partial y \end{bmatrix} = \begin{bmatrix} 1-y & -x \\ y & x \end{bmatrix}$$

giving $|J^{-1}| = x$. Hence $|J| = 1/x$ and the required integral is

$$\int_1^3\int_1^2 x^2 y(1/x)\,\mathrm{d}\alpha\,\mathrm{d}\beta = \int_1^3\int_1^2 xy\,\mathrm{d}\alpha\,\mathrm{d}\beta = \int_1^3\int_1^2 \beta\,\mathrm{d}\alpha\,\mathrm{d}\beta = 4$$

1.4 From symmetry the approximation must be of the form

$$\int_{-1}^1 f(\alpha)\,\mathrm{d}\alpha = h_1 f(-\alpha_1) + h_2 f(0) + h_1 f(\alpha_1)$$

This equation is always exact for any *odd* function, i.e. $f(\alpha) = \alpha, \alpha^3, \alpha^5, \ldots$, since both sides are always zero. If it is also true for the *even* functions $f(\alpha) = 1, \alpha^2, \alpha^4$ then

$$\int_{-1}^1 \mathrm{d}\alpha = 2 = 2h_1 + h_2, \quad \int_{-1}^1 \alpha^2\,\mathrm{d}\alpha = \tfrac{2}{3} = 2h_1\alpha_1^2, \quad \int_{-1}^1 \alpha^4\,\mathrm{d}\alpha = \tfrac{2}{5} = 2h_1\alpha_1^4$$

whence $\alpha_1 = \sqrt{\tfrac{3}{5}}, h_1 = \tfrac{5}{9}, h_2 = \tfrac{8}{9}$.

1.5 If only one approximating function is used, equation (1.11) reduces to a single equation with $k_{11} = H/3$ and $w_1 = W/12$. Hence

Fig. 1.11. Hermite polynomials for the interval $0 \leqslant x \leqslant 1$.

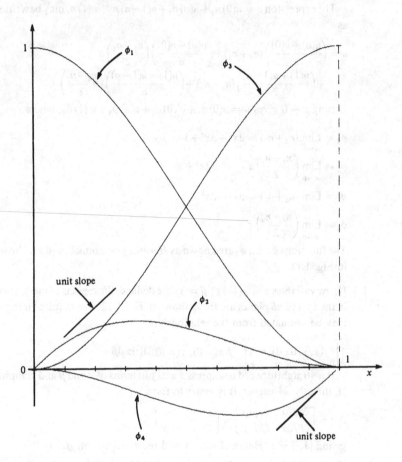

Fig. 1.12.

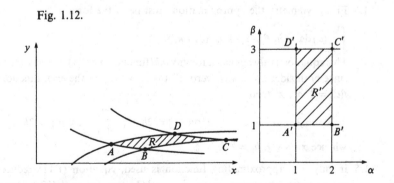

$c_1 = w_1/k_{11} = W/4H$. The approximate solution is thus $u = (W/4H)x(1-x)$, with

$$T(u) = \int_0^1 H(u')^2/2 - Wxu\,dx = -(\tfrac{1}{12})(W^2/8H)$$

If u is replaced by the true solution $\tilde{u} = Wx(1-x^2)/6H$ given in section 1.2, rather tedious integration gives $T(\tilde{u}) = -(\tfrac{4}{45})(W^2/8H)$, which is less than $T(u)$.

1.6 The *essential* boundary condition is $u = 0$ at $x = 0$. If the approximate solution does not satisfy this condition then the rod can move bodily and no minimum energy state exists. The *natural* boundary condition is $u' = \rho g L/E$ at $x = L$. The total potential energy for an approximation $u(x)$ is

$$T(u) = \int_0^L EA(u')^2/2 - \rho g\,Au\,dx - \rho g\,ALu(L)$$

Putting $\alpha = x/L$ gives the corresponding expression for the function $u(\alpha)$

$$T(u) = L\int_0^1 EA(u')^2/2L^2 - \rho g\,Au\,d\alpha - \rho gAL\,u(1)$$

The Ritz procedure may be illustrated by setting $u = [c_1 \sin(\pi\alpha/2) + c_2 \sin\pi\alpha](2L^2\rho g/E)$. Substitution of this approximation into the expression for $T(u)$ and differentiation with respect to c_1 and c_2 gives

$$\begin{bmatrix} \tfrac{1}{4}\pi^2 & 2\pi/3 \\ 2\pi/3 & \pi^2 \end{bmatrix}\begin{bmatrix} c_1 \\ c_2 \end{bmatrix} = \begin{bmatrix} 2/\pi + 1 \\ 2/\pi \end{bmatrix}$$

whence $c_1 = 0.742$, $c_2 = 0.093$. The exact and approximate solutions are shown in Fig. 1.13.

1.7 The Ritz equations may be obtained by substituting the approximation $u = c_i\phi_i$ into the expression (1.14) for the total potential energy (with $EI = 1$) and minimising $T(u)$ with respect to the coefficients c_i. Alterna-

Fig. 1.13.

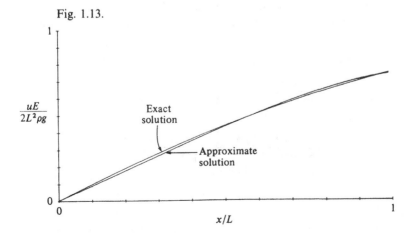

tively it is possible to start with the Galerkin equations (1.26)

$$\left[\int_0^1 \phi_j''' \, \phi_i \, dx\right] c_j = \int_0^1 \phi_i \, dx$$

and transform these equations into the required form using integration by parts. The procedure is very similar to that used in the case of the suspended cable problem discussed in section 1.5, except that two integrations are required.

First approximation: $\phi_2 = x^2$. There is only one Ritz equation

$$\int_0^1 4c_2 - x^2 \, dx = 0$$

giving $c_2 = \frac{1}{12}$ and $u = x^2/12$. This satisfies the essential boundary conditions $u(0) = u'(0) = 0$ but not the natural boundary condition $u''(1) = 0$ or the differential equation. The value of T_1 is $-\frac{1}{72}$.

Second approximation: $\phi_2 = x^2$, $\phi_3 = x^3$. There are two Ritz equations

$$\int_0^1 2(2c_2 + 6xc_3) - x^2 \, dx = 0 \qquad \left\{ \begin{matrix} 4c_2 + 6c_3 = \frac{1}{3} \\ 6c_2 + 12c_3 = \frac{1}{4} \end{matrix} \right.$$
$$\int_0^1 6x(2c_2 + 6xc_3) - x^3 \, dx = 0$$

giving $c_2 = \frac{5}{24}$, $c_3 = -\frac{1}{12}$. The associated solution satisfies the essential boundary conditions at $x = 0$ but not the natural boundary conditions at $x = 1$ or the differential equation. The value of T_2 is $-\frac{7}{288}$.

Third approximation: $\phi_2 = x^2$, $\phi_3 = x^3$, $\phi_4 = x^4$. There are three Ritz equations

$$4c_2 + 6c_3 + 8c_4 = \frac{1}{3}$$
$$6c_2 + 12c_3 + 18c_4 = \frac{1}{4}$$
$$8c_2 + 18c_3 + \frac{144}{5}c_4 = \frac{1}{5}$$

giving $c_2 = \frac{1}{4}$, $c_3 = -\frac{1}{6}$, $c_4 = \frac{1}{24}$ and $u = (6x^2 - 4x^3 + x^4)/24$. This solution satisfies the differential equation and all the boundary conditions and is therefore the true solution. The value of T_3 is $-\frac{1}{40}$.

Comparing the three approximations, $-\frac{1}{72} > -\frac{7}{288} > -\frac{1}{40}$, as expected.

Fig. 1.14.

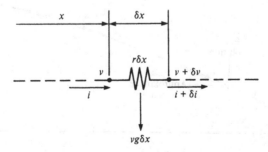

1.8 From Fig. 1.14 it follows that $\delta i = -vg\,\delta x$ and $\delta v = -ir\,\delta x$. Thus $d^2v/dx^2 = -r\,di/dx = rgv$, as required.

Let $\tilde{v}$ be the true solution and let $v = \tilde{v} + \varepsilon$ be an arbitrary distribution of potential satisfying the correct end conditions – i.e. $\varepsilon(0) = \varepsilon(L) = 0$. With a potential distribution v the power dissipated is

$$P(v) = \int_0^L (v')^2/r + gv^2\,dx$$
$$= \int_0^L (\tilde{v}')^2/r + g\tilde{v}^2\,dx + 2\int_0^L \tilde{v}'\varepsilon'/r + g\tilde{v}\varepsilon\,dx + \int_0^L (\varepsilon')^2/r + g\varepsilon^2\,dx$$

Integration by parts reduces the second integral to

$$2\left[\tilde{v}'\varepsilon/r \Big|_0^L + \int_0^L (-\tilde{v}''/r + g\tilde{v})\varepsilon\,dx \right]$$

The first term of this expression is zero since $\varepsilon(0) = \varepsilon(L) = 0$. The second term is zero since $\tilde{v}$ is the true solution and therefore satisfies the differential equation. Hence

$$P(v) = P(\tilde{v}) + \int_0^L (\varepsilon')^2/r + g(\varepsilon)^2\,dx$$

Clearly $P(v)$ is greater than $P(\tilde{v})$.

1.9 With the nodes numbered as shown in Fig. 1.10 the admittance or stiffness matrix takes the form

$$
\begin{bmatrix}
X & X & 0 & 0 & 0 & 0 & 0 & X & X & 0 & 0 & 0 & 0 & 0 & 0 & 0 \\
  & X & X & 0 & 0 & 0 & 0 & 0 & 0 & X & 0 & 0 & 0 & 0 & 0 & 0 \\
  &   & X & X & 0 & 0 & 0 & 0 & 0 & 0 & X & 0 & 0 & 0 & 0 & 0 \\
  &   &   & X & X & 0 & 0 & 0 & 0 & 0 & 0 & X & 0 & 0 & 0 & 0 \\
  &   &   &   & X & X & 0 & 0 & 0 & 0 & 0 & 0 & X & 0 & 0 & 0 \\
  &   &   &   &   & X & X & 0 & 0 & 0 & 0 & 0 & 0 & X & 0 & 0 \\
  &   &   &   &   &   & X & X & 0 & 0 & 0 & 0 & 0 & 0 & X & 0 \\
  &   &   & \text{symmetric} &   &   &   & X & X & 0 & 0 & 0 & 0 & 0 & 0 & X \\
  &   &   &   &   &   &   &   & X & X & 0 & 0 & 0 & 0 & 0 & X \\
  &   &   &   &   &   &   &   &   & X & X & 0 & 0 & 0 & 0 & 0 \\
  &   &   &   &   &   &   &   &   &   & X & X & 0 & 0 & 0 & 0 \\
  &   &   &   &   &   &   &   &   &   &   & X & X & 0 & 0 & 0 \\
  &   &   &   &   &   &   &   &   &   &   &   & X & X & 0 & 0 \\
  &   &   &   &   &   &   &   &   &   &   &   &   & X & X & 0 \\
  &   &   &   &   &   &   &   &   &   &   &   &   &   & X & X \\
  &   &   &   &   &   &   &   &   &   &   &   &   &   &   & X
\end{bmatrix}
$$

The *bandwidth*, equal to $1 + \text{Max}\,(i-j)$, where i, j are the node numbers at the ends of the elements, is 9.

If the nodes are re-numbered as follows

Old numbering   1 2 3 4 5 6 7 8 9 10 11 12 13 14 15 16
(Fig. 1.10)

New numbering   1 5 9 13 15 11 7 3 2   6 10 14 16 12 8 4

32   *1. Preliminaries*

the matrix takes the form

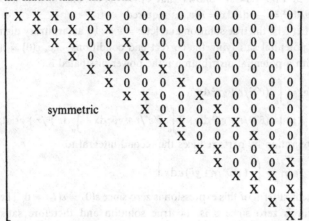

$$
\begin{bmatrix}
X & X & X & 0 & X & 0 & 0 & 0 & 0 & 0 & 0 & 0 & 0 & 0 & 0 & 0 \\
  & X & 0 & X & 0 & X & 0 & 0 & 0 & 0 & 0 & 0 & 0 & 0 & 0 & 0 \\
  &   & X & X & 0 & 0 & X & 0 & 0 & 0 & 0 & 0 & 0 & 0 & 0 & 0 \\
  &   &   & X & 0 & 0 & 0 & X & 0 & 0 & 0 & 0 & 0 & 0 & 0 & 0 \\
  &   &   &   & X & X & 0 & 0 & X & 0 & 0 & 0 & 0 & 0 & 0 & 0 \\
  &   &   &   &   & X & 0 & 0 & 0 & X & 0 & 0 & 0 & 0 & 0 & 0 \\
  &   &   &   &   &   & X & X & 0 & 0 & X & 0 & 0 & 0 & 0 & 0 \\
  & \text{symmetric} & & & & & & X & 0 & 0 & 0 & X & 0 & 0 & 0 & 0 \\
  &   &   &   &   &   &   &   & X & X & 0 & 0 & X & 0 & 0 & 0 \\
  &   &   &   &   &   &   &   &   & X & 0 & 0 & 0 & X & 0 & 0 \\
  &   &   &   &   &   &   &   &   &   & X & X & 0 & 0 & X & 0 \\
  &   &   &   &   &   &   &   &   &   &   & X & 0 & 0 & 0 & X \\
  &   &   &   &   &   &   &   &   &   &   &   & X & X & X & 0 \\
  &   &   &   &   &   &   &   &   &   &   &   &   & X & 0 & X \\
  &   &   &   &   &   &   &   &   &   &   &   &   &   & X & X \\
  &   &   &   &   &   &   &   &   &   &   &   &   &   &   & X
\end{bmatrix}
$$

with a bandwidth of 5.

2

The finite-element method introduced

While a large part of the analysis associated with the finite-element method is independent of the particular partial differential equation being solved, it is convenient to introduce the method in its simplest useful form – that is, in the context of the field problems associated with the steady flow of heat or electric current.

The analysis of these problems starts from three basic equations, which hold at all points within the region R for which a solution is required.

(1) The equation defining the potential or temperature gradient

$$\mathbf{e} = \nabla u \tag{2.1}$$

In this equation u is a temperature or electrical potential and $\mathbf{e}$ is the corresponding gradient vector.

(2) The equation defining the properties of the material

$$\mathbf{q} = -D\mathbf{e} \tag{2.2}$$

This equation relates $\mathbf{q}$, the density of heat or current flow, to the gradient of the temperature or electrical potential. Throughout this chapter it will be assumed that the conductivity D is a property of the material and is independent of $\mathbf{e}$ – i.e. the material is *linear*. (The symbol D is used to provide continuity of notation with later chapters of this book.)

(3) The equation of flow balance

$$\nabla \cdot \mathbf{q} = w \tag{2.3}$$

In this equation† w is a distribution of heat or current sources

---

† As mentioned in section 1.1, the use of the symbol $\nabla \cdot$ for the divergence operator implies the use of rectangular Cartesian coordinates.

within the body of the material and is assumed to be a known function of position.

Combining these three equations gives Poisson's equation

$$D\nabla^2 u = -w \tag{2.4}$$

which reduces to Laplace's equation if w is everywhere zero. Equations (2.1) to (2.4) hold for both two- and three-dimensional conduction problems.

Poisson's equation also governs the behaviour of a number of strictly two-dimensional phenomena, including
  (a) the elastic torsion of a shaft of arbitrary uniform cross-section,
  (b) the flow of viscous fluid in a pipe of arbitrary uniform cross-section,
  (c) the small deflection of an initially plane membrane under transverse loading. This is simply the two-dimensional version of the suspended-cable problem discussed in section 1.2.

A solution of equation (2.4) by the Ritz method described in section 1.2 begins with the specification of a set of M approximating functions ϕ_i within the solution region R, each function satisfying the essential boundary conditions (i.e. conditions involving the values of u on the boundary). An approximate solution $u = c_i \phi_i$ is then defined and finally the coefficients c_i are found by minimising the functional corresponding to (2.4). The accuracy of the solution is improved by increasing the number of approximating functions.

The finite-element method appears to follow a different approach, although this difference is more apparent than real, as will be plain later. The method divides the solution region R into sub-regions or 'finite elements', and defines an approximation u within each element, appropriate continuity conditions being imposed on the inter-element boundaries. Improvement in accuracy is achieved either by decreasing the size (and increasing the number) of the elements, or by increasing the number of terms in the approximations within the elements.

Much of the early finite-element literature treated continuum problems in the spirit of electric network or structural frame analysis, the finite elements being thought of as initially separate fragments of material, physically joined together by the analyst to form a continuous system. Later developments of the method require a more formal mathematical approach, but it is worth beginning by thinking of the method as a process of physical assembly.

2.1 The two-dimensional Poisson equation – a physical approach

As an illustrative example, consider the problem of finding the steady-state temperature distribution u in the electrical heating bar whose uniform cross-section is shown in Fig. 2.1a. Current flowing along the bar generates heat at a rate w per unit volume. This rate is not necessarily constant over the cross-section, but the distribution of w is assumed to be known. The surface of the bar is maintained at zero temperature and the thermal conductivity D is constant throughout the material.

A simple finite-element approach to this problem begins by taking a slice of the bar of unit thickness and dividing it into a set of triangular elements, such as the ones shown in Fig. 2.1b. The division of the cross-section into triangles means that the curved segments of boundary are replaced by a series of straight lines, but it is clear that increasing the number of triangles will decrease the error arising from this change, and that in the limit the true boundary S of R can be represented exactly.

A typical triangular element is shown shaded in Fig. 2.1b. The true

Fig. 2.1. (a) A problem of plane steady-state temperature distribution. (b) The solution region divided into triangular elements.

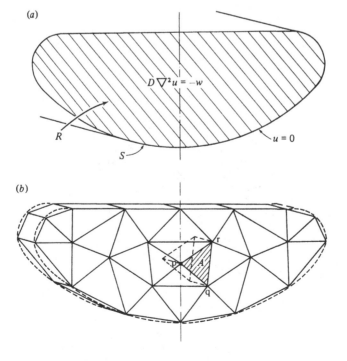

(a)

$D \nabla^2 u = -w$

R

S

$u = 0$

(b)

unknown temperature distribution $\tilde{u}$ is replaced by an approximation u which is linear within the triangle. This can be written in a form similar to (1.1) as

$$u(\mathbf{r}) = c_0 + \mathbf{c}_1 \cdot \mathbf{r}$$

where $\mathbf{r}$ is the position vector of an arbitrary point and $c_0, \mathbf{c}_1$ are constants. Alternatively u can be expressed in terms of its values u_1, u_2, u_3 at the vertices of the triangle, taking a form similar to (1.2), as

$$u(\mathbf{r}) = u_1 n_1(\mathbf{r}) + u_2 n_2(\mathbf{r}) + u_3 n_3(\mathbf{r}) = u_j n_j(\mathbf{r}) \quad (j = 1, 2, 3) \qquad (2.5)$$

where the functions n_j are linear functions of position or 'shape functions' of the type described in section 1.1, with the property that n_j takes the value 1 at vertex j and is zero at the other two vertices, for $j = 1, 2, 3$. It is straightforward to work out explicit expressions for the functions n_j from the geometry of Fig. 2.2, although these expressions are not actually needed in the present analysis.

The assumption that u is linear within the triangle means that the associated heat flow density $\mathbf{q}$ is constant. From (2.1), (2.2) and (2.5)

$$\mathbf{q} = -D\nabla u = -D(\nabla n_1 u_1 + \nabla n_2 u_2 + \nabla n_3 u_3) = -D\nabla n_j u_j \qquad (2.6)$$

Fig. 2.2. A typical triangular element: the quantities $\mathbf{n}_1, \mathbf{n}_2, \mathbf{n}_3$ are unit vectors normal to the sides of the triangle.

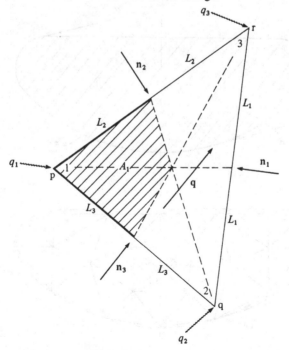

Since n_1 has value 1 at vertex 1 and value 0 at vertices 2 and 3 it follows from the geometry of the element that $\nabla n_1 = L_1 \mathbf{n}_1 / A$, where A is the area of the triangle and L_1 and $\mathbf{n}_1$ are defined in Fig. 2.2. Similarly $\nabla n_2 = L_2 \mathbf{n}_2 / A$, $\nabla n_3 = L_3 \mathbf{n}_3 / A$. Thus equation (2.6) may be written as

$$\mathbf{q} = -D(\mathbf{b}_1 u_1 + \mathbf{b}_2 u_2 + \mathbf{b}_3 u_3) = -D \mathbf{b}_j u_j \tag{2.7}$$

where $\mathbf{b}_1 = \nabla n_1 = L_1 \mathbf{n}_1 / A$, etc. Note that $|\mathbf{b}_j|$ is the reciprocal of the perpendicular distance from node j to the opposite side of the triangle.

The constant flow density $\mathbf{q}$ defined by (2.7) implies flows across the sides of the triangle. The next step is to replace these distributed flows by 'equivalent' concentrated inflows q_1, q_2, q_3 at the vertices. This replacement is achieved by dividing the inflow across each side equally between the vertices at the ends of the side. Thus half the inflow across the portion of boundary 3–1–2 in Fig. 2.2, equal in value to $(L_2 \mathbf{n}_2 + L_3 \mathbf{n}_3) \cdot \mathbf{q}$, is replaced by a concentrated inflow q_1 at vertex 1. Since $(L_2 \mathbf{n}_2 + L_3 \mathbf{n}_3) \cdot \mathbf{q} = -L_1 \mathbf{n}_1 \cdot \mathbf{q} = -A \mathbf{b}_1 \cdot \mathbf{q}$, etc., the three inflows are

$$q_i = -A \mathbf{b}_i \cdot \mathbf{q} \quad (i = 1, 2, 3) \tag{2.8}$$

Substituting for $\mathbf{q}$ from equation (2.7) gives $q_i = DA \, \mathbf{b}_i \cdot \mathbf{b}_j u_j$, which may be written in the form

$$\begin{bmatrix} q_1 \\ q_2 \\ q_3 \end{bmatrix} = \begin{bmatrix} k_{11} & k_{12} & k_{13} \\ & k_{22} & k_{23} \\ \text{symmetric} & & k_{33} \end{bmatrix} \begin{bmatrix} u_1 \\ u_2 \\ u_3 \end{bmatrix} \tag{2.9}$$

where $k_{ij} = DA \, \mathbf{b}_i \cdot \mathbf{b}_j$. The nodal flow/nodal temperature equations (2.9) for the three-noded triangle are singular, like the corresponding equations (1.32) for the two-noded resistor. A similar set of equations holds for each triangle in Fig. 2.1b.

Just as the distributed inflows across the sides of a triangle can be replaced by 'equivalent' concentrated inflows at its vertices, so the distribution of heat sources within the triangle can be replaced by 'equivalent' concentrated sources w_1, w_2, w_3 at the vertices. There are a number of ways of calculating the quantities w_i. The simplest way is to divide the triangle into three equal areas A_1, A_2, A_3 by means of the medians, as shown in Fig. 2.2. The total source strength within each of these areas is then regarded as being concentrated at the associated vertex, i.e.

$$w_i = \int_{A_i} w \, dR \quad (i = 1, 2, 3) \tag{2.10}$$

If w is constant over the triangle then each w_i is equal to $wA/3$.

The triangles are now connected together at their vertices to form the region R, in very much the same way as the resistors were connected to form a network in section 1.7. The process of connection associates a single value of u with each node – an unknown value if the node is internal to R and a known (zero) value if it is on the boundary S. Let there be M internal nodes. For the element shown in Fig. 2.2 the process identifies the vertices 1, 2, 3 of the isolated element with nodes p, q, r in the assembled system, so that u_1, u_2, u_3 are identical to u_p, u_q, u_r respectively. Since the function u is linear on each triangle edge, the continuity of u at the nodes implies continuity in the value of u across each inter-element boundary. Such a function is referred to as *piecewise-linear*.

The analysis of the assembly of elements now follows a very similar pattern to that of the assembly of resistors in section 1.7. The nodal current-balance equation (1.33) becomes the nodal flow-balance equation

$$\sum \frac{\text{equivalent concentrated nodal inflows to}}{\text{elements connected to node p}} = w_p \qquad (2.11)$$

where w_p is the sum of the equivalent concentrated sources at node p. Substituting from (2.9) into (2.11) for each element connected to node p gives a linear equation relating w_p (assumed known) to the values of u at node p and its immediate neighbours.† A similar procedure applied at each internal node of R gives a non-singular set of equations which can be solved for the M unknown nodal values u_m.‡ The contribution which the element with nodes p, q, r makes to these equations is

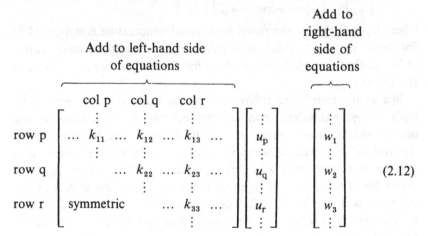

$$ (2.12) $$

---

† The fact that the equation associated with node p only involves values of u at p and its immediate neighbours means that the complete set of equations contains a large proportion of zeros. If the nodes are numbered in an ordered pattern the non-zero terms tend to be grouped round the leading diagonal. The computational implications of this are discussed in chapter 8.

‡ Note that the subscript m scans over the M internal nodes, while the subscripts i and j refer to a particular element and take only the values 1, 2, 3.

where dots indicate zero contributions. The coefficients in (2.12) should be compared with those in (1.34). The complete set of equations for the nodal variables u_m is simply the summation of the contributions from all the elements which make up R. Elements which have nodes on the boundary (where $u = 0$) contribute fewer terms to the final equations, since there are no rows or columns in (2.12) associated with boundary nodes. For example, if node 3 of the element in Fig. 2.2 is on the boundary, so that $u_3 = u_r = 0$, equation (2.9) reduces to

$$\begin{bmatrix} q_1 \\ q_2 \end{bmatrix} = \begin{bmatrix} k_{11} & k_{12} \\ \text{symmetric} & k_{22} \end{bmatrix} \begin{bmatrix} u_1 \\ u_2 \end{bmatrix}$$

(The expression for the flow q_3 is not required in the subsequent assembly process, since there are no flow-balance equations associated with boundary nodes.) The contribution of the element to the final set of equations is therefore

$$
\begin{array}{c}
\begin{array}{cc}
\text{Add to left-hand} & \text{Add to} \\
\text{side of equations} & \begin{array}{c}\text{right-hand}\\\text{side of}\\\text{equations}\end{array}
\end{array}\\[2pt]
\overbrace{}\quad\overbrace{}\\
\begin{array}{c}
\begin{array}{cc}\text{col } p & \text{col } q\end{array}\\
\begin{array}{cc}
\begin{array}{c}\text{row } p\\[20pt]\text{row } q\end{array}
\begin{bmatrix}
\vdots & \vdots \\
\cdots\; k_{11}\; \cdots & k_{12}\; \cdots \\
\vdots & \vdots \\
& \cdots\; k_{22}\; \cdots \\
\text{symmetric} & \vdots
\end{bmatrix}
\begin{bmatrix} \vdots \\ u_p \\ \vdots \\ u_q \\ \vdots \end{bmatrix}
&
\begin{bmatrix} \vdots \\ w_1 \\ \vdots \\ w_2 \\ \vdots \end{bmatrix}
\end{array}
\end{array}
\end{array}
\tag{2.13}
$$

The solution of the equations assembled in accordance with (2.12) and (2.13) gives an approximate solution u continuous in value throughout R and linear within each triangle, with a corresponding approximate flow distribution $\mathbf{q}$ which is constant within each triangle. This flow distribution satisfies $\boldsymbol{\nabla}\cdot\mathbf{q} = 0$ rather than $\boldsymbol{\nabla}\cdot\mathbf{q} = w$ within each triangle, while on each inter-element boundary there is, in general, a finite discontinuity in the normal component of $\mathbf{q}$, which corresponds to a 'line source' or a 'line sink', as shown in Fig. 2.3. However, although equation (2.3) is not satisfied, even approximately, at any *point* within R, it is satisfied in an *integral* sense over an area round each node. The area for node p is shown in Fig. 2.4, the associated integral condition being†

$$\int_{A_p} \boldsymbol{\nabla}\cdot\mathbf{q} - w\,\mathrm{d}R = 0$$

† This condition is proved in section 2.3.

This gives at least a plausible argument that reducing the size of the triangles will cause u to tend to the true solution. For as the triangles reduce in size, so the areas such as A_p over which flow balance is satisfied become smaller. As the area round each node shrinks, the integral condition satisfied by the approximate solution becomes closer to the point condition (2.3) satisfied everywhere by the true solution.

2.2   The two-dimensional Poisson equation – a Ritz approach

The approach of the previous section has a strong intuitive appeal, especially to anyone who comes to finite-element analysis from the analysis of electric networks or structural frames. The same problem can also be approached from a different point of view, which shows the finite-element method to be just a Ritz procedure with a particular choice of approximating functions. This allows the convergence of the method with decreasing element size to be established from the general convergence conditions associated with the Ritz process, as discussed in section 1.3.

The first step in this approach is the introduction of a set of *M nodal* shape functions N_m. These differ from the *element* shape functions introduced in the previous section in that each function is associated with a node in the connected assembly of elements and is defined throughout

Fig. 2.3. (*a*) A line source.   (*b*) A line sink.

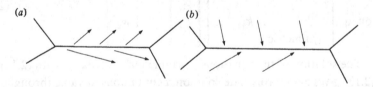

(*a*)   (*b*)

Fig. 2.4. The area of flow-balance for node p.

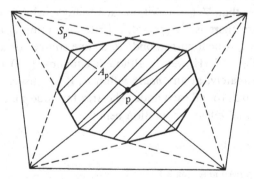

R. As in section 2.1, the subscript m scans over the M internal nodes of R. If p is a node within R, N_p is defined as the function which takes the value 1 at p, is zero at all other nodes and is linear within each triangular element of R. This function is shown in Fig. 2.5. Note that each function N_m is continuous in value within R and satisfies the condition $N_m = 0$ on the boundary S. It is therefore an admissible function as far as the Ritz process is concerned. Although the functions N_m are not orthogonal over R they possess the property that, within any triangle, there are at most three functions which are non-zero. For example, within the triangle with vertices p, q, r, only N_p, N_q and N_r are non-zero. The piecewise-linear approximation u introduced in the previous section may be written in the simple form

$$u = u_m N_m \quad (m = 1, ..., M) \tag{2.14}$$

The exact solution of equation (2.4) is that function u which, of all functions U satisfying the condition $U = 0$ on S, minimises the functional

$$T(U) = \int_R D(\nabla U)^2/2 - wU \, dR \tag{2.15}$$

In many problems (though not in heat conduction) the functional $T(U)$ has physical significance. For example, if U is the transverse displacement of a membrane stretched over an area R under constant tension D per unit length and carrying a transverse loading w per unit area, then the expression for $T(U)$ given in (2.15) is the total potential energy of the system.

Following the Ritz procedure described in section 1.2, the approximation $u = u_m N_m$ is substituted into (2.15), giving an integral which is a function of the M nodal variables u_m. Minimising $T(u)$ with respect to each of the

Fig. 2.5. The function N_p.

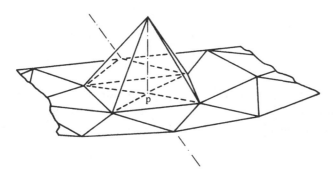

nodal variables gives a set of M equations

$$\frac{\partial T}{\partial u_l} = 0 = \int_R D \frac{\partial (\nabla u)}{\partial u_l} \cdot \nabla u - w \frac{\partial u}{\partial u_l} \, dR \quad (l = 1, ..., M) \qquad (2.16)$$

From (2.14) it follows that $\partial u / \partial u_l = N_l$ and $\partial (\nabla u) / \partial u_l = \nabla (\partial u / \partial u_l) = \nabla N_l$. Hence (2.16) becomes

$$\int_R D \nabla N_l \cdot \nabla u \, dR = \int_R w N_l \, dR \qquad (2.17)$$

or, since D is constant and $\nabla u = \nabla N_m u_m$

$$\left[D \int_R \nabla N_l \cdot \nabla N_m \, dR \right] u_m = \int_R w N_l \, dR \quad (l, m = 1, ..., M) \quad (2.18)$$

Equation (2.18) represents a set of M linear algebraic equations which can be solved for the nodal variables u_m. It should be compared with equation (1.10).

Now consider the contribution which the triangle with nodes p, q, r in Fig. 2.2 makes to the integrals in (2.18). Within this triangle the only non-zero shape functions are N_p, N_q, N_r, so that the triangle contributes to the complete set of equations (2.18) the terms

Left-hand side,
column m Right-hand side

$$\text{To row } l \qquad \left[D \int_A \nabla N_l \cdot \nabla N_m \, dA \right] u_m \qquad \int_A w N_l \, dA \qquad (2.19)$$

where A is the area of the triangle and l and m take only the values p, q, r.

Within A the shape functions N_p, N_q, N_r are identical to the functions n_1, n_2, n_3 defined in the previous section, so that $\nabla N_p = \nabla n_1 = \mathbf{b}_1$, etc. Hence (2.19) may be written as

Left-hand side,
column j' Right-hand side

$$\text{Add to row } i' \qquad \left[D \int_A \mathbf{b}_i \cdot \mathbf{b}_j \, dA \right] u_{j'} \qquad \int_A w n_i \, dA \qquad (2.20a)$$

where i, j take the values 1, 2, 3 and i', j' indicate the *global* node numbers associated with the *local* node numbers i, j. Since the vectors $\mathbf{b}_i$ are constant within A the left-hand integrals in (2.20a) reduce to simple products. Thus

(2.20a) may be written as

$$
\begin{array}{c}
\text{Add to left-hand side} \\
\text{of equations}
\end{array}
\qquad
\begin{array}{c}
\text{Add to} \\
\text{right-hand} \\
\text{side of} \\
\text{equations}
\end{array}
$$

$$
\begin{array}{c}
\text{row p} \\
\\
\text{row q} \\
\\
\text{row r}
\end{array}
\begin{bmatrix}
\cdots\ k_{11}\ \cdots\ k_{12}\ \cdots\ k_{13}\ \cdots \\
\\
\ \cdots\ k_{22}\ \cdots\ k_{23}\ \cdots \\
\\
\text{symmetric}\quad \cdots\ k_{33}\ \cdots
\end{bmatrix}
\begin{bmatrix}
u_{\mathrm{p}} \\
\\
u_{\mathrm{q}} \\
\\
u_{\mathrm{r}}
\end{bmatrix}
\begin{bmatrix}
w_1 \\
\\
w_2 \\
\\
w_3
\end{bmatrix}
\qquad (2.20b)
$$

where

$$
k_{ij} = DA\,\mathbf{b}_i\cdot\mathbf{b}_j \quad \text{and} \quad w_i = \int_A wn_i\,\mathrm{d}A \quad (i,j = 1,2,3) \qquad (2.21)
$$

Expressions (2.20b) and (2.12) are identical except for the difference in the definitions of the quantities w_i (compare equations (2.10) and (2.21)).

If w is constant over the element then (2.21), like (2.10), gives $w_i = wA/3$, since each of the functions n_i takes the form of a tetrahedron of unit height and base area A. In this case therefore, expressions (2.12) and (2.20b) are identical. This establishes the convergence of the procedure described in section 2.1 with decreasing element size, at least for the case where w is constant within each triangle.

If the element has one of its nodes on the boundary, expression (2.20) is modified in exactly the same way as expression (2.12).

2.3 The two approaches compared – non-uniform source distributions

If the quantities D and w in (2.2) and (2.3) are constant then the physical assembly process of section 2.1 produces the same equations for the nodal variables u_m as the Ritz process described in section 2.2. If D or w vary within an element then the equations produced by the two methods are slightly different. The nature of the difference is best seen by converting both sets of equations into a form which is essentially the integral condition (1.27) given in section 1.5. For Poisson's equation this means that the unknowns u_m are determined by the M equations

$$\int_R (\mathbf{\nabla} \cdot \mathbf{q} - w)\,\gamma_m\,dR = 0 \quad (m = 1, ..., M) \tag{2.22}$$

where the functions γ_m are a set of M independent functions specified within R. [The appearance of the term $\mathbf{\nabla} \cdot \mathbf{q}$ in (2.22) requires some comment, since if u is piecewise-linear then $\mathbf{q}$ has discontinuities of value on the inter-element boundaries. This implies that there are points within R where $\mathbf{\nabla} \cdot \mathbf{q}$ is not defined. This difficulty may be avoided by inserting 'boundary-layers' of thickness ε between the elements and letting the functions N_m vary smoothly through these layers in the same way as the function ϕ_s in Fig. 1.4b. This device makes the functions $\mathbf{\nabla} N_m$ and $\mathbf{\nabla} \cdot \mathbf{q}$ continuous throughout R, and ensures that the integrals in the analysis which follows are 'well-behaved'. It is straightforward to show that the contributions to (2.22) from the 'boundary layers' remain finite as $\varepsilon \to 0$.]

Consider first the procedure set out in section 2.1. Equation (2.11) for flow-balance at node p may be written as

$$\int_{S_p} \mathbf{q} \cdot \mathbf{n}\,dS = \int_{A_p} w\,dR \tag{2.23}$$

where A_p is the area enclosing node p in Fig. 2.4, S_p is the boundary of that area and $\mathbf{n}$ is the unit outward normal to the boundary. Application of the divergence theorem to the left-hand side of (2.23) gives

$$\int_{A_p} (\mathbf{\nabla} \cdot \mathbf{q} - w)\,dR = 0$$

which can be written as

$$\int_R (\mathbf{\nabla} \cdot \mathbf{q} - w)\,\bar{N}_p\,dR = 0 \tag{2.24}$$

where $\bar{N}_p$ is the 'top-hat' function shown in Fig. 2.6, whose value is 1 within A_p and 0 elsewhere. The complete set of equations set up by the

Fig. 2.6. The function $\bar{N}_p$.

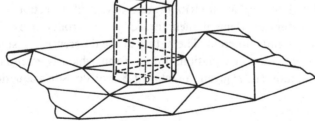

discretisation process of section 2.1 can thus be written in the form

$$\int_R (\boldsymbol{\nabla}\cdot\mathbf{q} - w)\, \bar{N}_m \, \mathrm{d}R = 0 \quad (m = 1, ..., M) \tag{2.25}$$

Now consider the Ritz procedure of section 2.2. In section 1.5 it was stated that the Ritz equations are identical to those set up by applying the Galerkin conditions. The proof given in that section for a simple one-dimensional example may easily be generalised to the two-dimensional Poisson equation. It is convenient to start the proof from the mathematical identity

$$\int_R \boldsymbol{\nabla}\cdot(N_m\mathbf{q})\, \mathrm{d}R = \int_R \boldsymbol{\nabla}N_m \cdot \mathbf{q}\, \mathrm{d}R + \int_R N_m \boldsymbol{\nabla}\cdot\mathbf{q}\, \mathrm{d}R \tag{2.26}$$

By the divergence theorem the left-hand side of (2.26) is equal to

$$\int_S N_m\mathbf{q}\cdot\mathbf{n}\, \mathrm{d}S$$

where $\mathbf{n}$ is the unit normal to the boundary S. This integral is zero, since N_m is zero everywhere on S for all values of m. It follows that

$$\int_R \boldsymbol{\nabla}N_m \cdot \mathbf{q}\, \mathrm{d}R = -\int_R N_m \boldsymbol{\nabla}\cdot\mathbf{q}\, \mathrm{d}R \tag{2.27}$$

Since $\mathbf{q} = -D\,\boldsymbol{\nabla}u$, (equation (2.6)) the Ritz equations (2.17) may be written in the form

$$\int_R \boldsymbol{\nabla}N_m \cdot \mathbf{q}\, \mathrm{d}R = -\int_R wN_m \, \mathrm{d}R \tag{2.28}$$

Combining equations (2.27) and (2.28) gives the Galerkin equations

$$\int_R (\boldsymbol{\nabla}\cdot\mathbf{q} - w)\, N_m \, \mathrm{d}R = 0 \quad (m = 1, ..., M) \tag{2.29}$$

It is now apparent that the difference between the procedures of sections 2.1 and 2.2 reduces to the difference between the functions $\bar{N}_m$ and N_m which appear in equations (2.25) and (2.29) respectively. Since

$$\int_R \bar{N}_p \, \mathrm{d}R = \int_R N_p \, \mathrm{d}R = \text{(Area of triangles surrounding p)}/3$$

equations (2.25) and (2.29) are identical if $\boldsymbol{\nabla}\cdot\mathbf{q}$ and w are constant within each triangle. It may be argued on physical grounds that if w varies, the Ritz process provides a better representation of the variation. Consider, for example, the representation of a concentrated source of strength W occurring within a triangle. In the physical approach of section 2.1 this source is simply transferred unchanged to the nearest node. In the Ritz process, on the other hand, it is replaced by 'equivalent' sources according to the 'lever rule' – it is as though W were a point mass, replaced by three statically equivalent point masses at the triangle vertices. This notion of equivalence is discussed further in section 4.7.

2.4   Finite elements and finite differences – a comparison

Consider now the case where w is constant throughout R and the elements in Fig. 2.1b are all equilateral triangles with sides of length h. Each vector $\mathbf{b}_i$ in (2.7) and (2.8) is now of magnitude $2/h\sqrt{3}$, while each triangle has area $A = h^2\sqrt{3}/4$. Hence in expressions (2.12) or (2.20)

$$k_{ij} = DA\,\mathbf{b}_i\cdot\mathbf{b}_j = D/\sqrt{3} \quad (i = j)$$
$$= -D/2\sqrt{3} \quad (i \neq j)$$
$$w_i = wA/3 = wh^2/4\sqrt{3}$$

Thus (2.20) becomes

(2.30)

The equation for node p in Fig. 2.7 is formed from contributions associated with the six elements which meet at that node, each contribution being similar to (2.30). Summing these contributions gives the equation

$$(D/\sqrt{3})[6u_\mathrm{p} - u_\mathrm{q}\ldots - u_\mathrm{v}] = 6wh^2/4\sqrt{3}$$

or

$$D(2/3h^2)[6u_\mathrm{p} - u_\mathrm{q}\ldots - u_\mathrm{v}] = w \qquad (2.31)$$

The right-hand side of (2.31) is exactly the same as the approximation for $-D\nabla^2 u$ generated from the finite-difference molecule given in Fig. 1.7c. Since a similar equation applies at all nodes in R, it follows that in this example the finite-element method and the finite-difference method generate exactly the same system of equations and therefore produce identical approximations to the true solution as far as the nodal values u_m are concerned.

Although the nodal values are the same, the two solutions cannot be regarded as completely identical. In the finite-element solution the nodal values are simply parameters associated with a piecewise-linear function u defined *throughout* the solution region. In the finite-difference solution the nodal points are the *only* points at which the solution is defined.

The two systems of equations also differ in the way in which the terms on the right-hand sides are calculated. If (2.31) is derived from a finite-element analysis, the right-hand side comes from an *averaging* process over a region surrounding node p. If w is not constant then the right-hand side must be determined by summing the results of (2.10) or (2.21) over the triangles round node p. If, however, (2.31) comes from a finite-difference approximation then the right-hand side is simply the *local* value of the function w at node p. The two right-hand sides will only be numerically equal if w is either constant or a linear function of position. (Equality occurs when w is a linear function because the mesh is uniform.)

2.5 Other boundary conditions

The analysis of sections 2.1 and 2.2 may easily be generalised to cover problems with more complicated boundary conditions.

An obvious extension is to the case where the solution is required to equal a specified function $\tilde{u}_S$ on the boundary S of the solution region. This condition can be satisfied approximately in a finite-element analysis by making the boundary nodal values u_s equal to the corresponding specified values $\tilde{u}_s$. In view of the fact that u is piecewise-linear throughout

Fig. 2.7. A node in a mesh of equilateral triangles.

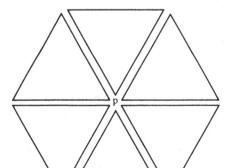

R this implies that, on the boundary S, u is equal to a piecewise-linear function u_S, which tends to $\tilde{u}_S$ as the mesh size is reduced.

The only change to the analysis presented in section 2.1 occurs in the assembly of the final set of equations for the unknown nodal variables u_m. Equation (2.9) still describes the properties of a general triangular element with vertices 1, 2, 3, but if vertex 3, say, is at boundary node r, at which the value of u is specified as $\tilde{u}_r$, then (2.9) must be written in the form

$$\begin{bmatrix} q_1 \\ q_2 \end{bmatrix} = \begin{bmatrix} k_{11} & k_{12} \\ \text{symmetric} & k_{22} \end{bmatrix} \begin{bmatrix} u_1 \\ u_2 \end{bmatrix} + \begin{bmatrix} k_{13} u_3 \\ k_{23} u_3 \end{bmatrix} \tag{2.32}$$

The contribution of the element to the complete set of equations is given by a modified version of (2.13) in which the terms $k_{13}\tilde{u}_r$ and $k_{23}\tilde{u}_r$ are subtracted from w_1 and w_2 respectively.

When viewed from the standpoint of the Ritz method the boundary condition $u = u_S$ is a non-homogeneous essential boundary condition. The approximating function must therefore take a form similar to (1.17), that is

$$u = u^{(0)} + u_m N_m \tag{2.33}$$

where $u^{(0)}$ is a function satisfying the non-homogeneous condition $u^{(0)} = u_S$ on S. This condition is satisfied by making $u^{(0)}$ equal to $\tilde{u}_s N_s$, where s scans over all the boundary nodes and N_s are the associated nodal shape functions. Note that the shape functions for the internal nodes N_m satisfy the homogeneous conditions $N_m = 0$ on S. Expression (2.33) clearly satisfies the required condition $u = u_S$ on S, whatever values are taken by the internal nodal variables u_m.

If (2.33) is used in section 2.2 in place of (2.14) equation (2.18) becomes

$$\left[D \int_R \nabla N_l \cdot \nabla N_m \, dR \right] u_m = \int_R w N_l \, dR - \left[D \int_R \nabla N_l \cdot \nabla N_s \, dR \right] \tilde{u}_s \tag{2.34}$$

The contribution of an element with nodes p, q, r is given by (2.19) with the right-hand-side term replaced by

$$\int_A w N_l \, dA - \left[D \int_A \nabla N_l \cdot \nabla N_s \, dA \right] \tilde{u}_s \tag{2.35}$$

If r is a boundary node then l takes only the values p and q, s takes only the value r and expression (2.35) gives exactly the same modification of (2.13) as that described above.

Another type of boundary condition is that in which the normal derivative rather than the value of u is specified. (It is not mathematically permissible to specify both value *and* normal derivative.) This is essentially a specification of the boundary inflow $\mathbf{q \cdot n}$ and is a 'natural' boundary condition. In the finite-element method the specified distribution of inflow is replaced by equivalent concentrated inflows at the boundary nodes, which are then treated as normal 'internal' nodes. Consider, for example, the heat-conduction problem posed in section 2.1 when the source distribution w is symmetric about the axis of symmetry of the cross-section. Under these conditions the solution region can be reduced to that shown in Fig. 2.8, with the condition $\mathbf{q \cdot n} = 0$ on the straight boundary AB. In the assembly of the nodal equations the two nodes on that boundary are treated as normal nodes with (in this particular example) zero external inflows.

The comments made in section 2.3 concerning the replacement of concentrated internal sources apply also to boundary inflows. The physical approach of section 2.1 merely transfers a concentrated boundary inflow to the nearest boundary node, while the Ritz/Galerkin procedure replaces it by inflows at the two adjacent nodes according to the 'lever rule'.

At least one node of a finite-element mesh must always be assigned a specified value of u. If this is not done, the solution will be able to 'float' and the nodal equations will inevitably be singular.

2.6 The three-dimensional Poisson equation

Only minor changes are needed to cover the more general case where R is a three-dimensional region. Consider for example, the analysis given in section 2.1. In three dimensions the triangular element of Fig. 2.2 is replaced by a tetrahedron, whose geometry is shown in Fig. 2.9. If u is

Fig. 2.8. Half a symmetrical solution region with zero inflow specified on the straight boundary AB.

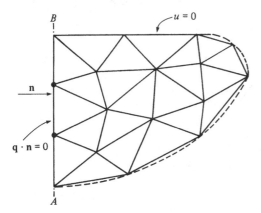

made linear within the tetrahedron then it can be expressed in the form $u = u_j n_j, j = 1...4$, where the shape functions n_j are again linear functions such that n_j takes the value 1 at vertex j and is zero at the other three vertices. Equation (2.6) remains unchanged as

$$\mathbf{q} = -D\nabla n_j u_j \tag{2.36}$$

and it follows from the geometry of the element that $\nabla n_1 = A_1 \mathbf{n}_1/3V$, etc., where V is the volume of the element. Thus (2.36) becomes

$$\mathbf{q} = -D\mathbf{b}_j u_j \tag{2.37}$$

where $\mathbf{b}_1 = A_1 \mathbf{n}_1/3V$, etc. Note that $|\mathbf{b}_j|$ is the reciprocal of the perpendicular distance from vertex j to the opposite face of the tetrahedron.

The remainder of the analysis in section 2.1 may be re-worked in a similar manner. Equation (2.9) becomes

$$\begin{bmatrix} q_1 \\ q_2 \\ q_3 \\ q_4 \end{bmatrix} = \begin{bmatrix} k_{11} & k_{12} & k_{13} & k_{14} \\ & k_{22} & k_{23} & k_{24} \\ & & k_{33} & k_{34} \\ \text{symmetric} & & & k_{44} \end{bmatrix} \begin{bmatrix} u_1 \\ u_2 \\ u_3 \\ u_4 \end{bmatrix} \tag{2.38}$$

where $k_{ij} = DV\mathbf{b}_i \cdot \mathbf{b}_j$. The equations which correspond to (2.10) and (2.21) are respectively

$$w_i = \int_{V_i} w\,dV \quad \text{and} \quad w_i = \int_V wn_i\,dV \quad (i = 1,...,4)$$

where V_i represents the four volumes into which the median planes divide V. If w is constant throughout the element volume then both these integrals are equal to $wV/4$ for all four values of i.

The complete set of equations for the nodal values u_j is built up in exactly the same way as before, the contribution of the element to the final

Fig. 2.9. A tetrahedral element: $\mathbf{n}_1$ is a unit vector normal to the face q r s.

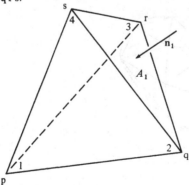

set of equations being given by (2.12), with the addition of a row and column to the matrix on the left-hand side and the corresponding addition of a term to the vector on the right. The analysis presented in other sections of this chapter can be modified in a similar manner.

2.7 Problems in which the conductivity varies

If the conductivity D is not constant throughout an element it cannot be taken outside the integral in equation (2.18). Consequently the coefficients k_{ij} in (2.20) take the more general form

$$k_{ij} = \int_A D\,\mathbf{b}_i \cdot \mathbf{b}_j\,\mathrm{d}A$$

where the integrals may have to be evaluated numerically. The three-dimensional analysis presented in section 2.6 may be modified in a similar manner.

Another important generalisation concerns cases where the conducting material is not isotropic. In such problems the conductivity D becomes a tensor rather than a scalar. This means that in a particular Cartesian coordinate system the conductivity is represented by a 2×2 or 3×3 symmetric matrix $\mathbf{D}$. The governing equations (2.1), (2.2) and (2.3) are now written in the form

$$\mathbf{e} = \nabla u, \quad \mathbf{q} = -\mathbf{D}\mathbf{e}, \quad \nabla^t \mathbf{q} = w \tag{2.39}$$

[Note that the scalar product $\nabla \cdot \mathbf{q}$ is written in the equivalent form $\nabla^t \mathbf{q}$ to make all three equations (2.39) conform to the rules of matrix algebra.]

The essential features of the analysis are not altered by this change, although care must be taken to preserve the sequence of symbols when manipulating expressions. Equation (2.15) must be written in the form

$$T(U) = \int_R (\nabla U)^t \mathbf{D} \nabla U/2 - wU\,\mathrm{d}R$$

with a corresponding change in (2.18),

$$\left[\int_R (\nabla N_l)^t \mathbf{D} \nabla N_m\,\mathrm{d}R \right] u_m = \int_R wN_l\,\mathrm{d}R \tag{2.40}$$

The coefficients k_{ij} in (2.20b) take the form

$$k_{ij} = \int_A \mathbf{b}_i^t \mathbf{D} \mathbf{b}_j\,\mathrm{d}A \quad \text{or} \quad k_{ij} = A\,\mathbf{b}_i^t \mathbf{D} \mathbf{b}_j$$

if $\mathbf{D}$ is constant within an element.

2.8   Some comments on the method

The nodal equations are in a sense the central equations of the finite-element method, and anyone writing a finite-element computer program will naturally think of the program in terms of (a) setting up and (b) solving these equations. However, it is a mistake to give the nodes too much importance – certainly a mistake to think that the solution u is likely to be most accurate at the nodes. The most important step in the method is really the definition of the piecewise approximation u – first by the division of the solution region R into sub-regions or finite elements, and then by the choice of a simple polynomial form within each element. The nodal variables u_m are simply the most convenient parameters for the definition of a function u with the correct degree of continuity.

The piecewise-linear form of u used in this chapter ensures that the satisfaction of continuity conditions at the nodes automatically leads to continuity of u across all the inter-element boundaries. As stated in section 1.3, this is a sufficient condition, as far as Poisson's equation is concerned, for the convergence of u to the true solution as the element size is reduced. Elements which provide sufficient continuity to satisfy the Ritz conditions of section 1.3 are said to 'conform', and much of the rest of this book is concerned with conforming elements. Note that the property of conforming is not just a characteristic of the functional approximation, but depends on the order of the differential equation being solved. Although finite-element analyses using non-conforming elements do not automatically converge to the correct solution, they can in practice be remarkably accurate, due to balancing of the errors resulting from finite mesh size on the one hand and boundary discontinuities on the other. Conditions for the correct convergence of such analyses are given in section 5.6.

Since the finite-element approach set out in this chapter is a Ritz procedure the inequality deduced in section 1.4 holds. Thus a finite-element analysis of a stretched membrane under a single load will underestimate the displacement of the load, while an analysis of heat or current flow from a single source will similarly underestimate the temperature or potential associated with that source. Note that 'single load' or 'single source' should be interpreted as a load or source distributed over a small but *finite* region. If the load or source is concentrated at a point then the deflection of the membrane or the potential at the source point becomes infinite.

The remainder of this book is concerned with extensions of the ideas presented in this chapter. These extensions take a variety of forms:

(a) The replacement of scalar quantities (e.g. temperatures or potentials) by vectors (e.g. displacements) and the replacement of vector quantities (e.g. flows) by tensors (e.g. stresses).

(b) The replacement of the simple differential operator ∇ by the more complex operators associated with stress analysis.

(c) The introduction of elements with more complicated geometrical boundaries – i.e. quadrilaterals, curvilinear triangles, etc.

(d) The introduction of higher order shape functions – i.e. elements in which the dependent variables can have quadratic or cubic variation.

The reader should not allow the notational changes associated with these extensions to obscure the fact that the basic pattern of the analysis remains unchanged.

Problems for chapter 2

2.1 Find the vectors $\mathbf{b}_i$ for the triangle whose nodes are (x_1, y_1), (x_2, y_2), (x_3, y_3).

2.2 Nodes 1 and 2 of a linear triangular element are on the boundary of the solution region. There is a distributed normal inflow across this section of the boundary which varies linearly from zero at node 1 to $2Q/L$ per unit length at node 2, where L is the distance from node 1 to node 2 and Q is the total inflow. Find the equivalent nodal inflows by (a) the approach of section 2.1, (b) the Ritz method described in section 2.2.

2.3 Carry out the equation-assembly process for a finite-element solution of $\nabla^2 u = 0$ in the two-dimensional region shown in Fig. 2.10, indicating zero coefficients in the nodal equations by a 0 and non-zero coefficients by a X.

2.4 A uniform conducting sheet occupies a two-dimensional region R and has a boundary S. The temperature on the boundary is prescribed as $u = \tilde{u}_S$. The sheet loses heat by radiation at a rate proportional to the local value of u, so that the steady-state differential equation satisfied by u is

$\nabla^2 u = k^2 u$

Show that the solution of this differential equation is that function which, of all functions U satisfying the condition $U = \tilde{u}_S$ on S, minimises the integral

$$\int_R (\nabla U)^2 + k^2 U^2 \, dR$$

(This problem is the two-dimensional analogue of problem 1.8.)

2.5 The longitudinal component of the electric field in a waveguide propagating in TM mode satisfies the Helmholtz equation

$$\nabla^2 E_z + k^2 E_z = 0$$

within the waveguide, with $E_z = 0$ on the boundary. Show that a finite-element analysis using linear triangular elements leads to a modified version of (2.19), in which the contribution to the nodal equations of a triangle with vertices at nodes p, q, r is given by

Add to column m

row l $\left[\int_A (\nabla N_l \cdot \nabla N_m - k^2 N_l N_m) \, dA \right] (E_z)_m$ $(l, m = p, q, r)$

the right-hand side of the equations being a column of zeros.

2.6 Repeat the analysis of problem 2.5 for the system described in problem 2.4. What is the significant difference between the solutions of the two sets of nodal equations?

2.7 Unsteady heat conduction in a two- or three-dimensional region satisfies the diffusion equation

$$D \nabla^2 u = c\rho \, \partial u / \partial t$$

Show that the finite-element procedure transforms this equation into a set of ordinary first-order differential equations in the nodal variables u_m.

Fig. 2.10.

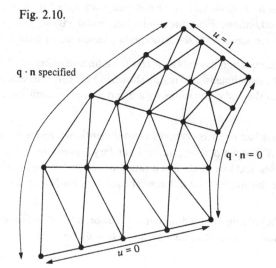

Solutions to problems

2.1 In section 2.1 $\mathbf{b}_1$ is defined as $\nabla n_1 = L_1 \mathbf{n}_1 / A$. From Fig. 2.11

$$\mathbf{b}_1 = (\tfrac{1}{2}A) \begin{bmatrix} -2L_1 \cos \psi \\ -2L_1 \sin \psi \end{bmatrix} = (\tfrac{1}{2}A) \begin{bmatrix} y_2 - y_3 \\ x_3 - x_2 \end{bmatrix}$$

where

$$2A = \begin{vmatrix} 1 & 1 & 1 \\ x_1 & x_2 & x_3 \\ y_1 & y_2 & y_3 \end{vmatrix}$$

The vectors $\mathbf{b}_2$ and $\mathbf{b}_3$ are obtained by appropriate permutation of the suffices.

2.2 The distribution of inflow is shown in Fig. 2.12, the inflow at distance x from node 1 being $q = 2Qx/L^2$ per unit length.

(a) The approach of section 2.1 replaces the distributed flow by the nodal inflows

$$q_1 = \int_0^{L/2} q \, dx = Q/4, \quad q_2 = \int_{L/2}^{L} q \, dx = 3Q/4$$

Fig. 2.11.

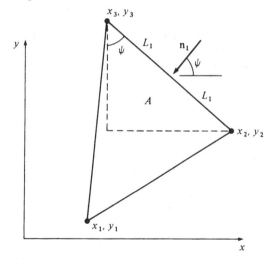

Fig. 2.12.

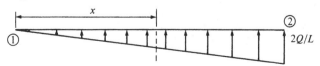

(b) The Ritz method of section 2.2 replaces the distributed flow by the nodal inflows

$$q_1 = \int_0^L n_1 q \, dx, \quad q_2 = \int_0^L n_2 q \, dx$$

where

$$n_1 = (L-x)/L, \quad n_2 = x/L, \quad \text{giving} \quad q_1 = Q/3, \quad q_2 = 2Q/3.$$

2.3 The arrangement of the equations depends on the numbering of the nodes. For the numbering shown in Fig. 2.13 the non-zero entries are

						Coefficients										Equivalent nodal inflows

```
┌X X 0 0 0 X 0 0 0 0 0 0 0 0 0┐   ┌X┐ w₁
│X X 0 0 X X 0 0 0 0 0 0 0 0 0│   │0│
│  X X 0 0 X X 0 0 0 0 0 0 0 0│   │0│
│    X X 0 0 X X 0 0 0 0 0 0 0│   │0│
│      X 0 0 0 X X 0 0 0 0 0 0│   │0│
│        X X 0 0 0 X 0 0 0 0 0│   │X│ w₆
│          X X 0 0 X X 0 0 0 0│   │0│
│ symmetric  X X 0 0 X X 0 0 0│   │0│
│              X X 0 0 X X 0 0│   │0│
│                X 0 0 0 X X  │   │0│
│                  X X 0 0 0  │   │X│ w₁₁ + terms due
│                    X X 0 0  │   │X│      to specified
│                      X X 0  │   │X│      value u = 1
│                        X X  │   │X│      on the
└                          X ┘   └X┘      boundary
```

2.4 The solution to this problem follows the same pattern as the solution already given for problem 1.8.

Let $\tilde{u}$ be the true solution of the differential equation with the specified

Fig. 2.13.

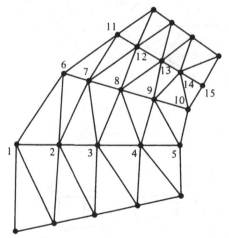

boundary condition and let $u = \tilde{u} + \varepsilon$ be a function satisfying the condition $u = \tilde{u}$ on S, i.e. $\varepsilon = 0$ on S. Then

$$T(u) - T(\tilde{u}) = \int_R (\nabla(\tilde{u} + \varepsilon))^2 - (\nabla\tilde{u})^2 + k^2(\tilde{u} + \varepsilon)^2 - k^2\tilde{u}^2 \, dR$$

$$= 2\int_R (\nabla\tilde{u} \cdot \nabla\varepsilon + k^2\tilde{u}\varepsilon) \, dR + \int_R (\nabla\varepsilon)^2 + k^2\varepsilon^2 \, dR \qquad (2.41)$$

Since $\nabla \cdot (\varepsilon\nabla\tilde{u}) = \nabla\varepsilon \cdot \nabla\tilde{u} + \varepsilon \nabla^2\tilde{u}$, the first integral in (2.41) may be written as

$$2\int_R \nabla \cdot (\varepsilon\nabla\tilde{u}) \, dR - 2\int_R \varepsilon(\nabla^2\tilde{u} - k^2\tilde{u}) \, dR$$

The first of these two integrals is zero since it may be transformed by the divergence theorem into $\int_S \varepsilon \nabla\tilde{u} \cdot \mathbf{n} \, dS$, where $\mathbf{n}$ is the unit normal to S, and ε is zero on S. The second of the two integrals is zero since $\tilde{u}$ is the true solution. Thus (2.41) may be written as

$$T(u) - T(\tilde{u}) = \int_R (\nabla\varepsilon)^2 + k^2\varepsilon^2 \, dR$$

which implies $T(u) > T(\tilde{u})$, as required.

2.5 From the solution to problem 2.4 the functional to be minimised is

$$T = \int_R (\nabla E_z)^2 - k^2 E_z^2 \, dR$$

If E_z is expressed in terms of a set of nodal values $(E_z)_m$ as $E_z = (E_z)_m N_m$ the procedure of section 2.2 gives

$$\frac{\partial T}{\partial (E_z)_l} = 0 = \int_R \frac{\partial(\nabla E_z)}{\partial(E_z)_l} \cdot \nabla E_z - \frac{k^2 \partial E_z}{\partial(E_z)_l} \cdot E_z \, dR \quad (l = 1, ..., M)$$

in place of equation (2.16) and

$$\left[\int_R \nabla N_l \cdot \nabla N_m - k^2 N_l N_m \, dR \right] (E_z)_m = 0 \quad (l, m = 1, ..., M) \qquad (2.42)$$

in place of (2.18). This gives the required result.

2.6 For the heat-conduction problem 2.4 the analysis is similar, with $-k^2$ replaced by $+k^2$ and non-zero terms appearing on the right-hand side of the equation (2.42) due to the specified values of $\tilde{u}$ on the boundary S.

However, the nature of the solution is quite different in the two cases. The Helmholtz equation gives rise to an eigenvalue problem: the nodal equations may be written in the form $[\mathbf{A} - k^2\mathbf{B}] \, \mathbf{E}_z = 0$, where the non-zero coefficients of $\mathbf{A}$ and $\mathbf{B}$ are arranged in the same pattern and $\mathbf{E}_z$ is the vector of nodal variables. Non-zero values of the nodal variables are only possible if k^2 takes certain values. The heat-conduction equation in problem 2.4, on the other hand, has a non-zero solution for any value of k^2.

2.7 The effect of replacing $-w$ in equation (2.4) by $c\rho\,\partial u/\partial t$ replaces the right-hand-side contribution in (2.19) by

$$-c\rho\int_A (\partial u/\partial t)\, N_l\, dA$$

Since $u = u_m N_m$ it follows that $\partial u/\partial t = (du_m/dt)\, N_m$. Hence the right-hand side of (2.19) becomes

$$-c\rho\left[\int_A N_m N_l\, dA\right](du_m/dt) \tag{2.43}$$

Thus the nodal equations take the form

$$\mathbf{A}u + c\rho\,\mathbf{B}(du/dt) = 0$$

where $\mathbf{A}$ and $\mathbf{B}$ are the same matrices as appear in the solution to problem 2.6 and $\mathbf{u}$ is the vector of nodal displacements u_m. This is a set of first-order differential equations in the nodal variables. The evaluation of integrals of the type appearing in (2.42) and (2.43) is discussed in the solution to problem 3.7.

3

Elastic stress analysis using linear triangular elements

This chapter develops the finite-element method in the context of two- and three-dimensional linear elastic stress analysis. The treatment is closely linked to that of chapter 2, using the same triangular and tetrahedral elements and the same piecewise-linear approximating functions.

The treatment of Poisson's equation in chapter 2 made use of the concepts and notation of vector calculus, so that there was no need to introduce any particular coordinate system. Although a similar approach is possible in the case of elastic stress analysis, the following presentation is based on a conventional x, y or x, y, z Cartesian coordinate system.

3.1 The equations of two-dimensional elasticity

In the Poisson problem of plane steady-state heat conduction the basic dependent variable is the *scalar* temperature u. In plane elasticity this variable is replaced by the *vector* displacement $\mathbf{u}$, with components u_x, u_y. In equation (2.1) the *vector* operator ∇ operates on u to produce the *vector* gradient $\mathbf{e}$. The analogous equation in plane elasticity involves a *tensor* operator, which operates on $\mathbf{u}$ to produce a strain *tensor*. However, if the subsequent analysis does not involve coordinate transformations, the operator and the strain tensor may be written as a matrix and a vector respectively. If this is done, and the two relationships set out side-by-side, the similarity between them becomes apparent,

plane Poisson problem   plane stress problem

$$\mathbf{e} = \begin{bmatrix} e_x \\ e_y \end{bmatrix} = \begin{bmatrix} \partial/\partial x \\ \partial/\partial y \end{bmatrix} u \quad \varepsilon = \begin{bmatrix} \varepsilon_{xx} \\ \varepsilon_{yy} \\ \gamma_{xy} \end{bmatrix} = \begin{bmatrix} \partial/\partial x & 0 \\ 0 & \partial/\partial y \\ \partial/\partial y & \partial/\partial x \end{bmatrix} \begin{bmatrix} u_x \\ u_y \end{bmatrix} \qquad (3.1a)$$

or

$$\mathbf{e} = \nabla u \qquad\qquad \varepsilon = \square\, \mathbf{u} \qquad\qquad (3.1b)$$

(Note that although the 'vector' ε formed from the strains does not transform as a true vector if the coordinate axes are rotated, it does satisfy the vector addition law.)

In equation (2.2) the flow $\mathbf{q}$ is related to the gradient $\mathbf{e}$ by the conductivity D. For isotropic material this quantity is a scalar, but it becomes a matrix $\mathbf{D}$ in the anisotropic case (see section 2.7). Similarly in plane elasticity the stress tensor is related to the strain tensor by a tensor dependent on the material properties. In a specified Cartesian coordinate system the stress tensor, like the strain tensor, can be written as a vector, while the tensor defining the material properties can be written as a matrix. The relationship for isotropic material under conditions of plane stress is

$$\begin{bmatrix} \sigma_{xx} \\ \sigma_{yy} \\ \tau_{xy} \end{bmatrix} = \frac{E}{1-v^2} \begin{bmatrix} 1 & v & 0 \\ v & 1 & 0 \\ 0 & 0 & (1-v)/2 \end{bmatrix} \begin{bmatrix} \varepsilon_{xx} \\ \varepsilon_{yy} \\ \gamma_{xy} \end{bmatrix} \qquad (3.2a)$$

where E is Young's modulus and v is Poisson's ratio. The same equations hold for problems of plane strain, with the constants E and v replaced by $E' = E/(1-v^2)$, and $v' = v/(1-v)$ respectively. Equations (2.2) and (3.2a) may be set out in comparable form as

$$\mathbf{q} = -D\,\mathbf{e} \qquad\qquad \sigma = \mathbf{D}\,\varepsilon \qquad\qquad (3.2b)$$

Finally, in equation (2.3) the flow $\mathbf{q}$ is required to have a certain divergence w. In plane elasticity the stress σ is required to satisfy an equilibrium condition, which involves a distributed body force $\mathbf{w}$ per unit area. Once again the two relationships may be set out side-by-side,

plane Poisson problem $\qquad\qquad$ plane stress problem

$$[\partial/\partial x \ \partial/\partial y] \begin{bmatrix} q_x \\ q_y \end{bmatrix} = w \qquad \begin{bmatrix} \partial/\partial x & 0 & \partial/\partial y \\ 0 & \partial/\partial y & \partial/\partial x \end{bmatrix} \begin{bmatrix} \sigma_{xx} \\ \sigma_{yy} \\ \tau_{xy} \end{bmatrix} = -\begin{bmatrix} w_x \\ w_y \end{bmatrix}$$

or $\qquad\qquad\qquad\qquad\qquad\qquad\qquad\qquad\qquad\qquad\qquad\qquad (3.3a)$

$$\nabla^t \mathbf{q} = w \qquad\qquad \square^t \sigma = -\mathbf{w} \qquad\qquad (3.3b)$$

The equilibrium equations (3.3a) may be derived by considering the equilibrium of the infinitesimal element shown in Fig. 3.1.

Boundary conditions correspond in a similar way. Over a given segment of boundary it is possible to specify either the displacement $\mathbf{u}$ or the 'traction', i.e. the distribution of normal and tangential boundary forces. It is not permissible to specify both 'displacement' *and* 'traction' conditions simultaneously. However, the vector nature of the boundary displacement and traction variables makes it possible to specify 'mixed' boundary conditions, in the sense that it is possible to specify, for example, a boundary with a zero normal component of displacement and zero shear traction. The specification of a boundary traction is equivalent to specification of the boundary strain components – i.e. the partial derivatives of **u**. In the terminology of section 1.3, specified displacements are *essential* boundary conditions, while specified tractions are *natural* boundary conditions.

3.2 A finite-element approach to two-dimensional stress analysis

Having established a notational correspondence between the variables of potential theory and plane stress analysis it would be possible to make a formal transposition of the analysis of the previous chapter to give the corresponding analysis for plane-stress finite elements. However, a re-development of the analysis in the new physical context may give a clearer understanding of the method. In this section the solution of equations (3.1*b*), (3.2*b*) and (3.3*b*) will be considered in the two-dimensional region *R* shown in Fig. 3.2*a*, with distributed loading **w** per unit volume and the displacement condition **u** = **0** on the portion of boundary S_1. The remainder of the boundary carries no external load and is allowed to displace freely.

In chapter 2 the finite-element method was first described in physical terms, using the idea of equivalent concentrated nodal flows. This was followed by the more abstract Ritz/Galerkin formulation. The same

Fig. 3.1. An infinitesimal element of area in a plane stress field.

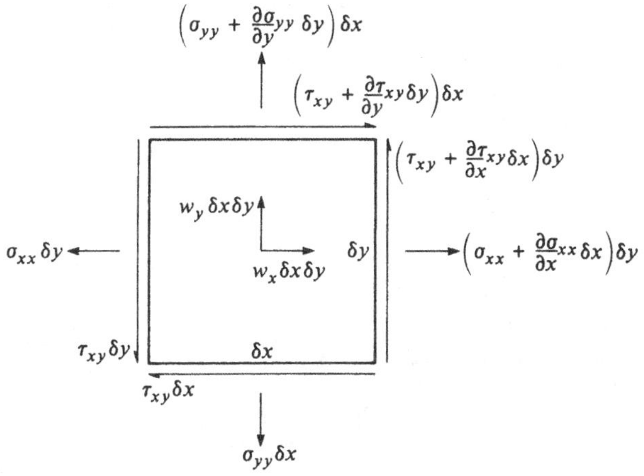

$$\left(\sigma_{yy} + \frac{\partial \sigma_{yy}}{\partial y} \delta y\right)\delta x$$

$$\left(\tau_{xy} + \frac{\partial \tau_{xy}}{\partial y}\delta y\right)\delta x$$

$$\left(\tau_{xy} + \frac{\partial \tau_{xy}}{\partial x}\delta x\right)\delta y$$

$w_y\,\delta x\delta y$

$\sigma_{xx}\,\delta y$

$w_x\,\delta x\delta y$

δy

$$\left(\sigma_{xx} + \frac{\partial \sigma_{xx}}{\partial x}\delta x\right)\delta y$$

$\tau_{xy}\delta y$

δx

$\tau_{xy}\delta x$

$\sigma_{yy}\delta x$

sequence could be followed in the case of plane stress analysis, beginning with a formulation based on equivalent concentrated nodal forces. Although this would be in accord with the historical development of the method, this formulation will be omitted here. As shown in the previous chapter, the Ritz process provides a more consistent treatment of what were there referred to as 'sources' – i.e. distributions of body and boundary forces in the present context, and provides a formal mathematical proof of the convergence of the method as the element size is reduced.

As in chapter 2, the region R is first divided into triangular elements, as shown in Fig. 3.2b. These elements can be thought of as being of unit thickness. The unknown vector displacement field is replaced by an approximation $\mathbf{u}$ which is linear within each triangle and continuous in value on the inter-element boundaries. Following the procedure of section 2.2, this approximation is written as

$$\mathbf{u}(x, y) = \mathbf{u}_m N_m(x, y) \quad (m = 1, ..., M) \tag{3.4}$$

where m scans over all the nodes of R at which displacements can occur, $\mathbf{u}_m$ represents the (unknown) displacements at those nodes and the functions N_m are the piecewise-linear nodal shape functions introduced in section 2.2 (see, for example, Fig. 2.5). It is worth stressing that these shape

Fig. 3.2. (*a*) A problem of plane stress distribution.
(*b*) The solution region divided into triangular elements.

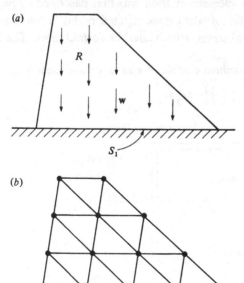

functions are quite independent of the scalar or vector nature of the dependent variable. They are determined solely by (a) the decision to use a piecewise-linear approximation for **u** and (b) the geometry of the triangles which make up R.

The piecewise-linear function **u** defined in (3.4) satisfies the conditions for the convergence of the Ritz process set out in section 1.3. Triangular elements with linear variation of displacement are therefore conforming elements as far as elastic stress analysis is concerned.

Corresponding to the displacement approximation **u** there is a distribution of strain ε, given by (3.1b) as $\varepsilon = \square\mathbf{u} = \square N_m \mathbf{u}_m$, and a distribution of stress σ, given by (3.2b) as

$$\sigma = \mathbf{D}\square\mathbf{u} = \mathbf{D}\square N_m \mathbf{u}_m \tag{3.5}$$

Note that since **u** is piecewise-linear, the functions ε and σ (like **e** and **q** in chapter 2), are constant within each element and have discontinuities of value on the inter-element boundaries.

In stress analysis it is usual to set up the finite-element equations by a virtual-work approach, rather than by minimisation of the total potential energy. As demonstrated in section 1.2, the principle of virtual work is simply a statement that the true solution is one of stationary potential energy, so that the analysis which follows is mathematically equivalent to the energy minimisation approach of section 2.2. The virtual displacement **u*** which is used in this analysis is restricted in the same way as the approximation **u**. That is, **u*** is chosen to be a piecewise-linear function, satisfying the condition $\mathbf{u}^* = \mathbf{0}$ on the boundary S_1 and thus expressible in terms of a set of nodal values $\mathbf{u}_l^*$ as

$$\mathbf{u}^* = \mathbf{u}_l^* N_l \quad (l = 1, \ldots, M) \tag{3.6}$$

In the virtual displacement **u*** the work done by the approximate stress system σ derived from **u** must equal that done by the applied loading **w**. This condition may be written as

$$\iint_R (\varepsilon^*)^t \sigma \, dx \, dy = \iint_R (\mathbf{u}^*)^t \mathbf{w} \, dx \, dy \tag{3.7}$$

where ε^* represents the distribution of virtual strains associated with the virtual displacements **u*** – i.e.

$$\varepsilon^* = \square\mathbf{u}^* = \square N_l \mathbf{u}_l^* \tag{3.8}$$

Substituting the expressions (3.5), (3.6) and (3.8) into (3.7) gives

$$\iint_R (\mathbf{u}_l^*)^t (\square N_l)^t \mathbf{D}\square N_m \mathbf{u}_m \, dx \, dy = \iint_R (\mathbf{u}_l^*)^t N_l \mathbf{w} \, dx \, dy$$

which may be written as

$$(\mathbf{u}_l^*)^t \iint_R (\Box N_l)^t \mathbf{D} \Box N_m \, dx \, dy \, \mathbf{u}_m = (\mathbf{u}_l^*)^t \iint_R N_l \mathbf{w} \, dx \, dy \qquad (3.9)$$

since the nodal displacements $\mathbf{u}_l^*$ and $\mathbf{u}_m$ are simply unknown multiplying constants as far as the integrations are concerned. As $\mathbf{u}_l^*$ represents an *arbitrary* set of virtual nodal displacements, equation (3.9) implies

$$\left[\iint_R (\Box N_l)^t \mathbf{D} \Box N_m \, dx \, dy \right] \mathbf{u}_m = \iint_R N_l \mathbf{w} \, dx \, dy \qquad (3.10)$$

Equation (3.10) represents a set of linear algebraic equations which can be solved for the nodal displacements $\mathbf{u}_m$. It should be compared with (2.18).

Now consider the contribution which the triangle with nodes p, q, r makes to the integrals in (3.10). Within the triangle the only shape functions which are non-zero are N_p, N_q, N_r, so that the triangle contributes the terms

Left-hand side, column m Right-hand side

$$\text{To row } l \quad \left[\iint_A (\Box N_l)^t \mathbf{D} \Box N_m \, dx \, dy \right] \mathbf{u}_m \quad \iint_A N_l \mathbf{w} \, dx \, dy \qquad (3.11)$$

where A is the area of the triangle and l, m take only the values p, q, r.

Within A the shape functions N_p, N_q, N_r are identical to the linear functions n_1, n_2, n_3 defined in section 2.1. From the definition of the operator $\Box$ in section 3.1 it follows that

$$\Box n_i = \begin{bmatrix} \partial n_i/\partial x & 0 \\ 0 & \partial n_i/\partial y \\ \partial n_i/\partial y & \partial n_i/\partial x \end{bmatrix} = \begin{bmatrix} (b_x)_i & 0 \\ 0 & (b_y)_i \\ (b_y)_i & (b_x)_i \end{bmatrix} \quad (i = 1, 2, 3) \qquad (3.12)$$

where $(b_x)_i$, $(b_y)_i$ are the components of the vectors $\mathbf{b}_i$ introduced in section 2.1. Remembering that $\mathbf{b}_1 = \mathbf{n}_1 L_1/A$, etc. it is easy to show from Fig. 3.3 that $(b_x)_1 = (y_2 - y_3)/2A$, and $(b_y)_1 = (x_3 - x_2)/2A$, etc.† so that

$$\Box n_1 = (\tfrac{1}{2}A) \begin{bmatrix} y_2 - y_3 & 0 \\ 0 & x_3 - x_2 \\ x_3 - x_2 & y_2 - y_3 \end{bmatrix}$$

with similar expressions for $\Box n_2$ and $\Box n_3$ by permutation of the suffices. By analogy with section 2.1, $\Box n_i$ is written as $\mathbf{B}_i$. The contribution which the triangle makes to the equations for the displacements $\mathbf{u}_m$ is therefore

Left-hand side,
column j' Right-hand side

$$\text{Add to row } i' \quad \left[\iint_A \mathbf{B}_i^t \mathbf{D} \mathbf{B}_j \, dx \, dy \right] \mathbf{u}_{j'} \quad \iint_A n_i \mathbf{w} \, dx \, dy \qquad (3.13a)$$

† See problem 2.1.

where i, j take the values 1, 2, 3 and i', j' indicate the *global* node numbers associated with the *local* node numbers i, j. The matrices $\mathbf{B}_i$, like the vectors $\mathbf{b}_i$ in sections 2.1 and 2.2, are constant within the element, so that the left-hand integrals in (3.13a) reduce to triple matrix products. Thus (3.13a) may be written as

$$
\begin{array}{c}
\overbrace{
\begin{array}{ccc}
\text{col p} & \text{col q} & \text{col r}
\end{array}
}^{\text{Add to left-hand side of equations}}
\qquad
\overbrace{}^{\substack{\text{Add to} \\ \text{right-hand} \\ \text{side of} \\ \text{equations}}}
\\[2pt]
\begin{array}{c}
\text{row p} \\[10pt]
\text{row q} \\[10pt]
\text{row r}
\end{array}
\left[
\begin{array}{ccc}
\cdots\ \mathbf{K}_{11}\ \cdots & \mathbf{K}_{12}\ \cdots & \mathbf{K}_{13}\ \cdots \\
& \mathbf{K}_{22}\ \cdots & \mathbf{K}_{23}\ \cdots \\
\text{symmetric} & & \mathbf{K}_{33}\ \cdots
\end{array}
\right]
\left[
\begin{array}{c}
\mathbf{u}_p \\ \mathbf{u}_q \\ \mathbf{u}_r
\end{array}
\right]
\left[
\begin{array}{c}
\mathbf{w}_1 \\ \mathbf{w}_2 \\ \mathbf{w}_3
\end{array}
\right]
\end{array}
\tag{3.13b}
$$

where

$$
\mathbf{K}_{ij} = \mathbf{B}_i^t \mathbf{D} \mathbf{B}_j A, \quad \mathbf{w}_i = \iint_A n_i \mathbf{w}\, \mathrm{d}x\, \mathrm{d}y \quad (i,j = 1, 2, 3)
\tag{3.14}
$$

Note that the terms 'row' and 'column' in (3.13) should be interpreted

Fig. 3.3. A typical triangular element.

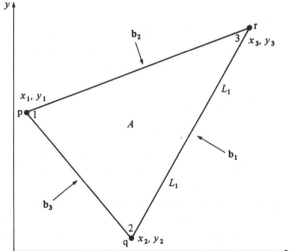

as references to the two scalar rows or columns associated with the two components of $\mathbf{u}$ at a node. If $\mathbf{w}$ is constant within the element then $\mathbf{w}_i = \mathbf{w}A/3$, as in section 2.2. The arrangement of coefficients in (3.13) should be compared with those in (2.20). The matrices $\mathbf{K}_{ij}$ are referred to as 'nodal stiffness matrices' and the vectors $\mathbf{w}_i$ as 'equivalent nodal loads'.

Nodes on the boundary S_1 are treated in the manner described in section 2.1. Thus if node 3 in Fig. 3.3 is on S_1 then $\mathbf{u}_r = 0$ and the terms associated with $\mathbf{u}_r$ are dropped from (3.13) to give expressions similar in arrangement to (2.13). Nodes on the remainder of the boundary are treated in the same way as internal nodes.

The assembly of the complete set of nodal load/displacement equations follows a similar pattern to that described in section 2.2, the only difference being the replacement of the coefficients k_{ij} by the 2×2 matrices $\mathbf{K}_{ij}$ and the corresponding replacement of the scalar source terms w_i by the vector load terms $\mathbf{w}_i$. (This difference corresponds to the difference between electrical network analysis and pin-jointed frame analysis set out in section 1.7). The set of nine $\mathbf{K}_{ij}$ matrices for a single element in (3.13b) forms a singular matrix, since an isolated element can be given an arbitrary rigid-body displacement. However, provided sufficient boundary nodes are fixed to prevent rigid-body movement of the whole continuum the assembled equations will be non-singular and can be solved for the nodal displacements $\mathbf{u}_m$. The displacement approximation $\mathbf{u}$ follows from (3.4), and the corresponding approximate strain and stress distributions can be found from equations (3.1) and (3.2).

3.3   Other boundary conditions

If a boundary node has a prescribed (non-zero) displacement, then it is treated in the manner already described for Poisson's equation in section 2.5. Thus if $\mathbf{u}_r$ in (3.13b) has a specified value $\tilde{\mathbf{u}}_r$ then the terms $\mathbf{K}_{13}\tilde{\mathbf{u}}_r$, $\mathbf{K}_{23}\tilde{\mathbf{u}}_r$ are evaluated and subtracted from the right-hand side of the equations. The two scalar equations associated with node r (row r in (3.13b)) are omitted.

If a portion of a boundary has known applied tractions (and unknown displacements) the boundary tractions must be replaced by equivalent concentrated loads at the boundary nodes. As with distributed inflows in Poisson's equation, it is possible simply to transfer external forces unchanged to the nearest boundary node, but it is better to make the concentrated nodal loads statically equivalent to the distributed loading by using the 'lever rule' of elementary mechanics. A more general technique for calculating equivalent nodal loads is described in section 4.7.

As mentioned earlier, the vector nature of displacements and loads makes it possible to have a boundary on which one scalar displacement condition and one scalar traction condition are specified. Consider, for example, the boundary shown in Fig. 3.4a, which has zero normal displacement and zero shear stress. In the finite-element approximation in Fig. 3.4b the equivalent conditions at node p are $\mathbf{u_p \cdot n} = 0$, $\mathbf{w_p \cdot t} = 0$, where $\mathbf{u_p}$ and $\mathbf{w_p}$ have unknown components in the $\mathbf{t}$ and $\mathbf{n}$ directions respectively.

One of the simplest ways of imposing these conditions is to add a series of very stiff pin-ended bars to the finite-element system, as shown in Fig. 3.4c. These prevent displacements normal to the boundary, but impose no constraints on tangential displacements. If the bars have axial stiffness $EA/L = K$ (chosen to be large compared with other stiffness coefficients appearing in the analysis) then the attachment of a bar to node p in the finite-element assembly implies the addition of a nodal stiffness matrix of magnitude

$$\mathbf{K_{pp}} = K \begin{bmatrix} \cos^2\psi & \cos\psi \sin\psi \\ \cos\psi \sin\psi & \sin^2\psi \end{bmatrix}$$

Fig. 3.4. (a) A boundary with zero shear traction and zero normal displacement.
(b) Boundary conditions at node p of the finite-element mesh.
(c) Pin-ended bars used to generate the required boundary conditions.

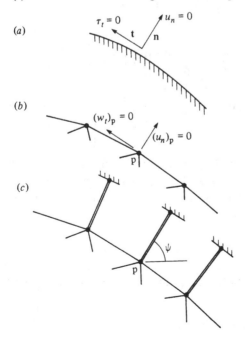

to the leading diagonal element associated with node p (i.e. row p, column p in (3.13b)). Once this has been done, the analysis proceeds with node p (and similar boundary nodes) treated as normal displacing nodes. Once the displacements have been found, the external forces at the boundary nodes may be computed from the nodal forces acting on the vertices of the associated elements.

When modelling complex boundary conditions it is important to remember that sufficient nodal displacement constraints must be provided to prevent rigid-body motion of the continuum.

3.4   Extending the analysis to three dimensions

The change to three dimensions follows a very similar pattern to that already presented in section 2.6. In rectangular Cartesian coordinates the displacement and body-force vectors $\mathbf{u}$ and $\mathbf{w}$ each have three components, while the strain and stress 'vectors' ε and σ each have six components. Equations (3.1a) and (3.2a) become

$$
\begin{bmatrix} \varepsilon_{xx} \\ \varepsilon_{yy} \\ \varepsilon_{zz} \\ \gamma_{xy} \\ \gamma_{yz} \\ \gamma_{zx} \end{bmatrix} = \begin{bmatrix} \partial/\partial x & 0 & 0 \\ 0 & \partial/\partial y & 0 \\ 0 & 0 & \partial/\partial z \\ \partial/\partial y & \partial/\partial x & 0 \\ 0 & \partial/\partial z & \partial/\partial y \\ \partial/\partial z & 0 & \partial/\partial x \end{bmatrix} \begin{bmatrix} u_x \\ u_y \\ u_z \end{bmatrix}
\tag{3.15a}
$$

and

$$
\begin{bmatrix} \sigma_{xx} \\ \sigma_{yy} \\ \sigma_{zz} \\ \tau_{xy} \\ \tau_{yz} \\ \tau_{zx} \end{bmatrix} = c_1 \begin{bmatrix} 1 & c_2 & c_2 & 0 & 0 & 0 \\ & 1 & c_2 & 0 & 0 & 0 \\ & & 1 & 0 & 0 & 0 \\ & \text{symmetric} & & c_3 & 0 & 0 \\ & & & & c_3 & 0 \\ & & & & & c_3 \end{bmatrix} \begin{bmatrix} \varepsilon_{xx} \\ \varepsilon_{yy} \\ \varepsilon_{zz} \\ \gamma_{xy} \\ \gamma_{yz} \\ \gamma_{zx} \end{bmatrix}
\tag{3.16a}
$$

where $c_1 = E(1-v)/(1+v)(1-2v)$, $c_2 = v/(1-v)$, $c_3 = (1-2v)/2(1-v)$. Re-defining the symbols ε, σ, $\mathbf{u}$, $\mathbf{w}$, $\mathbf{D}$ and $\square$ to represent three-dimensional quantities allows (3.15a) and (3.16a) to be written in the same form as (3.1b) and (3.2b)

$$
\varepsilon = \square \mathbf{u}
\tag{3.15b}
$$

$$
\sigma = \mathbf{D}\varepsilon
\tag{3.16b}
$$

With the re-defined symbols the equilibrium equation may be written as

$$
\square^t \sigma = -\mathbf{w}
\tag{3.17}
$$

which is identical in form to (3.3b).

The three-dimensional equivalent of the triangle is the tetrahedral element shown in Fig. 2.9. The displacement approximation within the element is written as $\mathbf{u} = \mathbf{u}_j n_j$, $(j = 1,...,4)$, so that $\boldsymbol{\sigma} = \mathbf{D}\boldsymbol{\varepsilon} = \mathbf{D}\square\mathbf{u} = \mathbf{D}\square n_j \mathbf{u}_j$, where the four nodal values $\mathbf{u}_j$ now each have three components and the shape functions n_j are exactly the same as the ones used in section 2.6. From the definition of the operator $\square$ in (3.15) it follows that

$$\square n_i = \begin{bmatrix} \partial n_i/\partial x & 0 & 0 \\ 0 & \partial n_i/\partial y & 0 \\ 0 & 0 & \partial n_i/\partial z \\ \partial n_i/\partial y & \partial n_i/\partial x & 0 \\ 0 & \partial n_i/\partial z & \partial n_i/\partial y \\ \partial n_i/\partial z & 0 & \partial n_i/\partial x \end{bmatrix} = \begin{bmatrix} (b_x)_i & 0 & 0 \\ 0 & (b_y)_i & 0 \\ 0 & 0 & (b_z)_i \\ (b_y)_i & (b_x)_i & 0 \\ 0 & (b_z)_i & (b_y)_i \\ (b_z)_i & 0 & (b_x)_i \end{bmatrix} \quad \substack{(i = 1, ..., 4) \\ \\ (3.18)}$$

where $(b_x)_i$, $(b_y)_i$, $(b_z)_i$ are the components of the four three-dimensional vectors $\mathbf{b}_i$ defined in section 2.6. (As noted in that section, each of these vectors is normal to a face of the tetrahedron and has magnitude equal to the reciprocal of the perpendicular from the face to the associated vertex.) As in the two-dimensional case, equation (3.18) is written as $\square n_i = \mathbf{B}_i$, the matrices $\mathbf{B}_i$ being constant throughout the element volume V.

A repetition of the analysis in section 3.2 results in a modified version of (3.13), in which an extra row and column, associated with the additional node s, appears in the coefficient matrix and load vector. The matrices $\mathbf{K}_{ij}$ are 3×3 matrices and the nodal load vectors $\mathbf{w}_i$ each have three components, the defining equations being

$$\mathbf{K}_{ij} = \mathbf{B}_i^t \mathbf{D} \mathbf{B}_j V, \quad \mathbf{w}_i = \iiint_V n_i \mathbf{w} \, dx \, dy \, dz \quad (i, j = 1, ..., 4) \quad (3.19)$$

If $\mathbf{w}$ is constant within the element then $\mathbf{w}_i = \mathbf{w}V/4$, as in section 2.6.

3.5 Further comments on the method

The linear displacement triangle and tetrahedron are often used in commercial stress-analysis, since irregularly-shaped boundaries are easily dealt with and different element sizes can be used in different parts of the solution region. Small elements can be used in the vicinity of corners and other points where stress gradients are likely to be large, while large elements can be used to represent areas of less interest, as shown in Fig. 3.5. The fact that the products $\mathbf{B}_i^t \mathbf{D} \mathbf{B}_j$ are constant within each element makes it easy to program the evaluation of the nodal stiffness matrices $\mathbf{K}_{ij}$.

In a solution based on linear triangular or tetrahedral elements there is continuity of displacement **u** throughout the solution region and the stress-strain conditions (3.2) or (3.16) are satisfied at all points. However, the stress distribution $\boldsymbol{\sigma}$ satisfies $\square^t\boldsymbol{\sigma} = \mathbf{0}$ rather than $\square^t\boldsymbol{\sigma} = \mathbf{w}$, with discontinuities of stress on the inter-element boundaries. The effect of the Ritz process described in section 3.2 is to make these discontinuities 'balance' within an area round each node, in the same way as the line sources and line sinks were made to balance in section 2.1 (The area associated with a node is given in Fig. 2.4). Thus the 'point' equilibrium condition (3.3) or (3.17) is replaced by an integral equilibrium condition. As in section 2.1, the area of integration decreases as the mesh size is reduced, which suggests that a piecewise-linear solution will converge to the correct solution. A formal proof, of course, comes from the general convergence properties of the Ritz process.

It is straightforward to generalise the analysis presented in section 1.4 to two- and three-dimensional stress analysis. The displacement condition (1.24) becomes

$$\int_R \mathbf{w}^t \mathbf{u}/2\,\mathrm{d}R \leqslant \int_R \mathbf{w}^t \bar{\mathbf{u}}/2\,\mathrm{d}R \tag{3.20}$$

where the two sides of the inequality represent the work done by the applied loads in the approximate and exact displacements. As mentioned

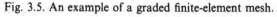

Fig. 3.5. An example of a graded finite-element mesh.

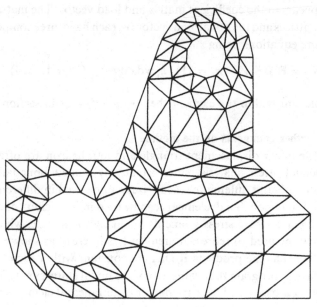

in section 1.4, this inequality is often translated into the statement that the method underestimates the displacement produced by a single point load. However, it should be remembered that a concentrated force at a point in an elastic continuum produces stresses and displacements which are infinite at the point of application of the load, the variation with the distance r from the point of application being as follows.†

	Two dimensions	Three dimensions
Stresses proportional to	$1/r$	$1/r^2$
Displacements proportional to	$\log r$	$1/r$

It is better to imagine the load distributed over a small but finite region of the continuum. Equation (3.20) may then be interpreted as stating that the *average* displacement of the loading zone in the direction of the applied load is underestimated by a finite-element solution using linear triangular or tetrahedral elements.

A numerical example showing the kind of accuracy which can be obtained with linear triangles is given in section 4.9. That section also compares the performance of linear elements with the higher-order elements introduced in the next chapter.

Problems for chapter 3

3.1 Prove that for isotropic material under conditions of plane *strain* $(\varepsilon_{zz} = 0)$ the stress-strain relationship is given by equations (3.2), with Young's modulus E and Poisson's ratio v replaced by $E' = E/(1-v^2)$ and $v' = v/(1-v)$ respectively.

3.2 A plane-stress triangular element of unit thickness has nodes $(0,0)$, $(1,0)$, $(0,1)$. Construct the shape functions n_i and the matrices $\mathbf{B}_i$. If $\mathbf{u}_i$ are the nodal displacements and $\mathbf{q}_i$ the corresponding equivalent nodal forces, verify the relationships $\varepsilon = \mathbf{B}_j \mathbf{u}_j$, $\mathbf{q}_i = A \mathbf{B}_i^t \sigma$ by direct physical reasoning.

3.3 In the plane-stress triangle shown in Fig. 3.6, node p lies on an axis of symmetry of the solution and nodes q and r are symmetrically disposed about that axis. What contribution, in terms of the $\mathbf{K}_{ij}$ matrices for the complete triangle, does the element make to the set of nodal equations in an analysis using nodes in the shaded region of the figure only?

† See, for example, reference 4, p. 97 (two-dimensional case) and p. 399 (three-dimensional case).

3.4 If thermal effects are included in the analysis of plane stress, equations (3.2) must be replaced by

$$\sigma = \mathbf{D}(\varepsilon - \varepsilon_\theta)$$

where $(\varepsilon_{xx})_\theta = (\varepsilon_{yy})_\theta = \alpha\theta$, $(\gamma_{xy})_\theta = 0$, θ is the temperature and α is the coefficient of thermal expansion. If the distribution of θ is given, show that the introduction of thermal effects results in the addition of terms

$$\iint_A \mathbf{B}_i^t \mathbf{D}\,\varepsilon_\theta\,\mathrm{d}x\,\mathrm{d}y$$

to the right-hand side of the nodal load/displacement equations (3.13b).

3.5 Use the result of problem 3.4 and the analysis of section 2.2 to construct the contribution which a single triangle makes to the nodal equations in a combined thermal-flow/elastic-stress analysis, in which the nodal variable at node i is $\begin{bmatrix} \mathbf{u}_i \\ \theta_i \end{bmatrix}$. Assume that both the displacement $\mathbf{u}_i$ and the temperature θ_i are piecewise-linear within the solution region.

3.6 If dynamic effects are included in the plane-stress analysis of section 3.2 the left-hand side of equations (3.13b) must be modified by the addition of acceleration terms $\mathbf{M}_{ij}(\mathrm{d}^2\mathbf{u}_{j'}/\mathrm{d}t^2)$. Find expressions for the 'mass matrices' $\mathbf{M}_{ij}$, (a) by assuming that the distributed mass ρA of each triangle is replaced by three masses, each of magnitude $\rho A/3$, at the vertices, (b) by assuming that the mass $\rho\,\mathrm{d}A$ of an elemental area $\mathrm{d}A$ imposes an inertia load $-\rho(\partial^2\mathbf{u}/\partial t^2)\,\mathrm{d}A$, where $\mathbf{u}$ is the usual piecewise-linear approximation to the displacement, and treating this loading in the same way as a normal static body-force distribution. (These two sets of matrices are termed *lumped* and *consistent* mass matrices respectively.)

3.7 Evaluate the mass matrices $\mathbf{M}_{ij}$ developed as the answer to problem 3.6b for a general plane triangle of constant density. [It is convenient to

Fig. 3.6. A triangular element situated on an axis of symmetry.

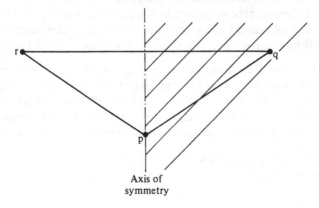

Axis of
symmetry

introduce a coordinate transformation which maps the general triangle
into the triangle $(0,0)$, $(1,0)$, $(0,1)$.]

Solutions to problems

3.1 The 'plane strain' version of **D** is most easily derived by setting
$\varepsilon_{zz} = \gamma_{yz} = \gamma_{zx} = 0$ in the full three-dimensional form (3.16a), giving

$$
\begin{bmatrix} \sigma_{xx} \\ \sigma_{yy} \\ \tau_{xy} \end{bmatrix} = E \begin{bmatrix} \dfrac{(1-v)}{(1+v)(1-2v)} & \dfrac{v}{(1+v)(1-2v)} & 0 \\ \dfrac{v}{(1+v)(1-2v)} & \dfrac{(1-v)}{(1+v)(1-2v)} & 0 \\ 0 & 0 & \dfrac{1}{2(1+v)} \end{bmatrix} \begin{bmatrix} \varepsilon_{xx} \\ \varepsilon_{yy} \\ \gamma_{xy} \end{bmatrix}
$$

It is a matter of simple algebra to verify that

$$
\left. \begin{array}{l} \dfrac{E(1-v)}{(1+v)(1-2v)} = \dfrac{E'}{1-v'^2} \\[2mm] \dfrac{Ev}{(1+v)(1-2v)} = \dfrac{E'v'}{1-v'^2} \\[2mm] \dfrac{E}{2(1+v)} = \dfrac{E'}{2(1+v')} \end{array} \right\} \begin{array}{l} \text{where} \\[2mm] E' = E/(1-v^2) \\[2mm] v' = v/(1-v) \end{array}
$$

giving

$$
\begin{bmatrix} \sigma_{xx} \\ \sigma_{yy} \\ \tau_{xy} \end{bmatrix} = \frac{E'}{1-v'^2} \begin{bmatrix} 1 & v' & 0 \\ v' & 1 & 0 \\ 0 & 0 & (1-v')/2 \end{bmatrix} \begin{bmatrix} \varepsilon_{xx} \\ \varepsilon_{yy} \\ \gamma_{xy} \end{bmatrix}
$$

which should be compared with equation (3.2a).

3.2 For this triangle the shape functions are $n_1 = 1-x-y$, $n_2 = x$, $n_3 = y$.
From equation (3.12)

$$
\mathbf{B}_1 = \begin{bmatrix} -1 & 0 \\ 0 & -1 \\ -1 & -1 \end{bmatrix}, \quad \mathbf{B}_2 = \begin{bmatrix} 1 & 0 \\ 0 & 0 \\ 0 & 1 \end{bmatrix}, \quad \mathbf{B}_3 = \begin{bmatrix} 0 & 0 \\ 0 & 1 \\ 1 & 0 \end{bmatrix}
$$

Hence

$$
\varepsilon = \mathbf{B}_j \mathbf{u}_j = \begin{bmatrix} -1 & 0 \\ 0 & -1 \\ -1 & -1 \end{bmatrix} \begin{bmatrix} u_{x1} \\ u_{y1} \end{bmatrix} + \begin{bmatrix} 1 & 0 \\ 0 & 0 \\ 0 & 1 \end{bmatrix} \begin{bmatrix} u_{x2} \\ u_{y2} \end{bmatrix} + \begin{bmatrix} 0 & 0 \\ 0 & 1 \\ 1 & 0 \end{bmatrix} \begin{bmatrix} u_{x3} \\ u_{y3} \end{bmatrix}
$$

giving

$$
\begin{aligned}
\varepsilon_{xx} &= u_{x2} - u_{x1} \\
\varepsilon_{yy} &= u_{y3} - u_{y1} \\
\gamma_{xy} &= (u_{y2} - u_{y1}) + (u_{x3} - u_{x1})
\end{aligned}
$$

which agrees with values obtainable by direct physical argument.

Substitution of the above expression for $\mathbf{B}_1$ into the equation $\mathbf{q}_1 = A\mathbf{B}_1^t\boldsymbol{\sigma}$ gives

$$\begin{bmatrix} q_{x1} \\ q_{y1} \end{bmatrix} = \tfrac{1}{2}\begin{bmatrix} -1 & 0 & -1 \\ 0 & -1 & -1 \end{bmatrix}\begin{bmatrix} \sigma_{xx} \\ \sigma_{yy} \\ \tau_{xy} \end{bmatrix}$$

i.e. $q_{x1} = -(\sigma_{xx}+\tau_{xy})/2$, $q_{y1} = -(\sigma_{yy}+\tau_{xy})/2$, which agrees with the result obtained from a direct equilibrium consideration of the heavy boundary shown in Fig. 3.7. Note that $\mathbf{q}_1$ is *equivalent* to the boundary tractions, not *in equilibrium* with them. A similar procedure may be carried out for nodes 2 and 3. These results should be compared with (2.8).

3.3 If symmetry is ignored, the contribution which the element makes to the complete set of nodal equations is given by (3.13*b*)

	Add to left-hand side of equations				Add to right-hand side of equations
	col p	col q	col r		
row p	$\cdots \mathbf{K}_{11} \cdots$	$\mathbf{K}_{12} \cdots$	$\mathbf{K}_{13} \cdots$	$\mathbf{u}_p$	$\mathbf{w}_1$
row q		$\cdots \mathbf{K}_{22} \cdots$	$\mathbf{K}_{23} \cdots$	$\mathbf{u}_q$	$\mathbf{w}_2$
row r	symmetric		$\cdots \mathbf{K}_{33} \cdots$	$\mathbf{u}_r$	$\mathbf{w}_3$

(3.21)

Symmetry imposes the condition $\begin{bmatrix} (u_x)_r \\ (u_y)_r \end{bmatrix} = \begin{bmatrix} -(u_x)_q \\ (u_y)_q \end{bmatrix}$, or $\mathbf{u}_r = \mathbf{C}\mathbf{u}_q$, where

Fig. 3.7.

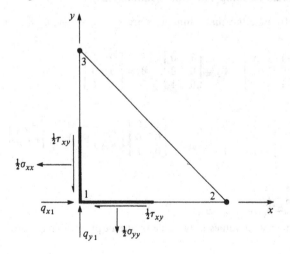

$C = \begin{bmatrix} -1 & 0 \\ 0 & 1 \end{bmatrix}$. Hence $\mathbf{u}_r$ may be eliminated from the equations associated with nodes p and q, reducing (3.21) to

	Add to left-hand side of equations		Add to right-hand side of equations
	col p	col q	

$$\text{row p} \begin{bmatrix} \cdots & \bar{\mathbf{K}}_{11} & \cdots & \mathbf{K}_{12} & \cdots \\ & \vdots & & \vdots & \\ \cdots & \bar{\mathbf{K}}_{21} & \cdots & \mathbf{K}_{22} & \cdots \end{bmatrix} \begin{bmatrix} \vdots \\ \mathbf{u}_p \\ \vdots \\ \mathbf{u}_q \\ \vdots \end{bmatrix} \qquad \begin{bmatrix} \vdots \\ \mathbf{w}_1 \\ \vdots \\ \mathbf{w}_2 \\ \vdots \end{bmatrix}$$

where $\bar{\mathbf{K}}_{11} = \mathbf{K}_{11} - \mathbf{K}_{13}\mathbf{C}$ and $\bar{\mathbf{K}}_{21} = \mathbf{K}_{21} - \mathbf{K}_{23}\mathbf{C}$. The equilibrium equation for node r is no longer required. Symmetry also imposes the condition $(u_x)_p = 0$. This condition can be dealt with by the method described in section 3.3. Note that the reduced set of equations is no longer symmetric.

3.4 If $\boldsymbol{\sigma} = \mathbf{D}(\boldsymbol{\varepsilon} - \boldsymbol{\varepsilon}_\theta)$ equation (3.7) becomes

$$\iint_R (\boldsymbol{\varepsilon}^*)^t \boldsymbol{\sigma} \, dx \, dy = \iint_R (\boldsymbol{\varepsilon}^*)^t \mathbf{D}(\boldsymbol{\varepsilon} - \boldsymbol{\varepsilon}_\theta) \, dx \, dy = \iint_R (\mathbf{u}^*)^t \mathbf{w} \, dx \, dy$$

If $\mathbf{u}^*$ is made equal to $\mathbf{u}_l^* N_l$, analysis similar to that of section (3.2) generates additional terms $\iint_A (\Box N_l)^t \mathbf{D}\boldsymbol{\varepsilon}_\theta \, dx \, dy$ on the right-hand side of (3.11). Since $\Box N_l = \mathbf{B}_i$, this is equivalent to the result given.

3.5 If $\boldsymbol{\varepsilon}_\theta = \begin{bmatrix} \alpha\theta \\ \alpha\theta \\ 0 \end{bmatrix} = \alpha \begin{bmatrix} 1 \\ 1 \\ 0 \end{bmatrix} \theta$ is written as $\alpha \begin{bmatrix} 1 \\ 1 \\ 0 \end{bmatrix} n_j \theta_{j'}$ ($j = 1, 2, 3; j' = \text{p}, \text{q}, \text{r}$) the

contribution to the nodal equilibrium equations due to thermal stress in a single triangle becomes

$$\mathbf{B}_i^t \mathbf{D}\alpha \begin{bmatrix} 1 \\ 1 \\ 0 \end{bmatrix} \iint_A n_j \, dx \, dy \; \theta_{j'}$$

where $\iint_A n_j \, dx \, dy = A/3$ for each value of j. Thus the contribution for $i = 1$ becomes

$$\frac{\alpha E}{6(1-v)} \begin{bmatrix} y_2 - y_3 \\ x_3 - x_2 \end{bmatrix} (\theta_p + \theta_q + \theta_r) = \frac{\alpha E}{6(1-v)} \mathbf{b}_1 (\theta_p + \theta_q + \theta_r)$$

with corresponding results for $i = 2, 3$. These additions to the nodal equilibrium equations may be written in a form similar to (3.13b) as

Add to right-hand side of equations

$$
\begin{array}{c}
\text{col p} \quad\; \text{col q} \quad\; \text{col r} \\
\begin{array}{c}
\text{row p}\\
\text{row q}\\
\text{row r}
\end{array}
\left[
\begin{array}{ccc}
\cdots\ \mathbf{c}_1\ \cdots & \mathbf{c}_1\ \cdots & \mathbf{c}_1\ \cdots \\
\cdots\ \mathbf{c}_2\ \cdots & \mathbf{c}_2\ \cdots & \mathbf{c}_2\ \cdots \\
\cdots\ \mathbf{c}_3\ \cdots & \mathbf{c}_3\ \cdots & \mathbf{c}_3\ \cdots
\end{array}
\right]
\left[
\begin{array}{c}
\theta_p\\
\theta_q\\
\theta_r
\end{array}
\right]
\end{array}
$$

where $\mathbf{c}_i = \alpha E \mathbf{b}_i/6(1-v)$.

The nodal temperatures $\theta_{j'}$ also satisfy the nodal heat-conduction equations (2.12). If the two sets of nodal equations are combined, the contribution of a single element becomes

Add to left-hand side of equations

Add to right-hand side of equations

$$
\begin{array}{c}
\text{col p} \quad\ \text{col q} \quad\ \text{col r} \\
\begin{array}{c}
\text{row p}\\
\text{row q}\\
\text{row r}
\end{array}
\left[
\begin{array}{ccc}
\cdots\ \bar{\bar{\mathbf{K}}}_{11}\ \cdots & \bar{\bar{\mathbf{K}}}_{12}\ \cdots & \bar{\bar{\mathbf{K}}}_{13}\ .. \\
\cdots\ \bar{\bar{\mathbf{K}}}_{21}\ \cdots & \bar{\bar{\mathbf{K}}}_{22}\ \cdots & \bar{\bar{\mathbf{K}}}_{23}\ .. \\
\cdots\ \bar{\bar{\mathbf{K}}}_{31}\ \cdots & \bar{\bar{\mathbf{K}}}_{32}\ \cdots & \bar{\bar{\mathbf{K}}}_{33}\ ..
\end{array}
\right]
\left[
\begin{array}{c}
\bar{\mathbf{u}}_p\\
\bar{\mathbf{u}}_p\\
\bar{\mathbf{u}}_r
\end{array}
\right]
\left[
\begin{array}{c}
\bar{\mathbf{w}}_1\\
\bar{\mathbf{w}}_2\\
\bar{\mathbf{w}}_3
\end{array}
\right]
\end{array}
$$

where

$$
\bar{\bar{\mathbf{K}}}_{ij} = \begin{bmatrix} \mathbf{K}_{ij} & -\mathbf{c}_i \\ 0 & k_{ij} \end{bmatrix}, \quad
\bar{\mathbf{u}}_{j'} = \begin{bmatrix} \mathbf{u}_{j'} \\ \theta_{j'} \end{bmatrix}, \quad
\bar{\mathbf{w}}_j = \begin{bmatrix} \mathbf{w}_j \\ w_j \end{bmatrix}
$$

and w_j represents the appropriate equivalent nodal heat inflows, as in (2.12). Each $\bar{\bar{\mathbf{K}}}_{ij}$ is a 3×3 matrix. Note that the nodal equations are no longer symmetric.

3.6 (*a*) If the distributed mass of a triangle is replaced by three masses $\rho A/3$ at the vertices, the acceleration of these masses produces terms $(\rho A/3)\ddot{\mathbf{u}}_{j'}$

Fig. 3.8.

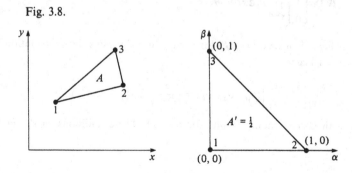

in the three nodal equations of motion. These contributions are added to the normal static stiffness terms in the nodal equations and may be written in the general form

Add to left-hand side of equations

$$
\begin{array}{c}
\text{col p} \quad \text{col q} \quad \text{col r} \\
\text{row p} \\
\text{row q} \\
\text{row r}
\end{array}
\begin{bmatrix}
\cdots \ \mathbf{M}_{11} \ \cdots & \mathbf{M}_{12} \ \cdots & \mathbf{M}_{13} \ \cdots \\
& \cdots \ \mathbf{M}_{22} \ \cdots & \mathbf{M}_{23} \ \cdots \\
\text{symmetric} & & \cdots \ \mathbf{M}_{33} \ \cdots
\end{bmatrix}
\begin{bmatrix}
\vdots \\ \ddot{\mathbf{u}}_p \\ \vdots \\ \ddot{\mathbf{u}}_q \\ \vdots \\ \ddot{\mathbf{u}}_r \\ \vdots
\end{bmatrix}
\tag{3.22}
$$

where $\mathbf{M}_{ij} = \begin{bmatrix} \rho A/3 & 0 \\ 0 & \rho A/3 \end{bmatrix}$ if $i = j$ and $\mathbf{0}$ if $i \neq j$.

(*b*) If the displacement **u** within a triangle takes the piecewise linear form $\mathbf{u} = \mathbf{u}_{j'} n_j$ then the acceleration $\ddot{\mathbf{u}}$ is given by $\ddot{\mathbf{u}} = \ddot{\mathbf{u}}_{j'} n_j$. The inertia loading is therefore $\mathbf{w} = -\rho \ddot{\mathbf{u}}_{j'} n_j$. Substitution of this expression for $\mathbf{w}$ into equation (3.14) gives nodal loading terms $-[\iint_A \rho n_i n_j \, dx \, dy] \ddot{\mathbf{u}}_{j'}$, implying mass matrices in equation (3.22) of the form

$$
\mathbf{M}_{ij} = \begin{bmatrix} \displaystyle\iint_A \rho n_i n_j \, dx \, dy & 0 \\ 0 & \displaystyle\iint_A \rho n_i n_j \, dx \, dy \end{bmatrix}
\tag{3.23}
$$

3.7 The consistent mass matrices developed in the solution to problem 3.6*b* require the evaluation of the integrals $\rho \iint_A n_i n_j \, dx \, dy$. Similar integrals also appeared earlier in the solutions to problems 2.5, 2.6 and 2.7. These integrals may be evaluated by using the mapping shown in Fig. 3.8. Transforming to variables α, β gives

$$
\iint_A n_i n_j \, dx \, dy = \iint_{A'} n_i(\alpha, \beta) \, n_j(\alpha, \beta) \, |J| \, d\alpha \, d\beta
$$

where $|J|$, the Jacobian of the mapping, is equal to $2A$, the ratio of the areas of the two triangles. Hence the diagonal elements of the mass matrices $\mathbf{M}_{ij}$ defined in (3.23) are equal to $2\rho A \displaystyle\int_0^1 \int_0^{1-\beta} n_i n_j \, d\alpha \, d\beta$, where

$n_1 = 1 - \alpha - \beta, \ n_2 = \alpha, \ n_3 = \beta.$ Hence $\mathbf{M}_{ij}$ is equal to $\begin{bmatrix} \rho A/6 & 0 \\ 0 & \rho A/6 \end{bmatrix}$ if $i = j$ and $\begin{bmatrix} \rho A/12 & 0 \\ 0 & \rho A/12 \end{bmatrix}$ if $i \neq j$. [Note that it is only necessary to evaluate the mass matrices for node 1 of the general triangle, since any node of this triangle may be mapped into node 1 of the triangle in the α, β plane.]

4

Higher-order approximations: (1) fixed element shapes

The simple piecewise-linear approximations used in chapters 2 and 3 involve linear shape functions which are easy to construct. Since triangular elements based on these functions make the dependent variable continuous in value, the associated finite-element solutions of Poisson's equation and the equations of linear elasticity converge to the true solutions as the element size is reduced. However, an approximation which assumes constant flow or stress in each element cannot model accurately situations in which these quantities vary rapidly without the use of a fine mesh, and this implies a large number of nodal variables. In such cases it is computationally more efficient to use elements in which the dependent variable can vary quadratically or cubically within each element, since fewer elements (and fewer nodal variables) are required for a given degree of accuracy.

In this chapter and the next, approximations are developed which are quadratic or cubic within the individual elements and continuous in value on the inter-element boundaries. The shape functions associated with these approximations can be used equally well in Poisson problems (u a scalar) or stress analysis problems ($\mathbf{u}$ a vector), since in both these cases conforming elements require continuity in the value of u or $\mathbf{u}$ but not in its derivatives.

Although the basic idea is a simple one, the explicit determination of higher-order shape functions for general triangles, quadrilaterals, etc. is algebraically very tedious. Much of this algebra can be avoided by dividing the work into two stages. The present chapter develops elements of fixed geometrical form, whose shape functions are easy to construct. The next chapter shows how these functions can also be used as geometrical mapping functions in the determination of the characteristics of more general elements. As in chapters 2 and 3, the emphasis is on two-dimensional elements, the extension of the ideas to three dimensions being indicated briefly.

The chapter concludes with a numerical example showing the accuracy obtainable with some of the elements discussed.

4.1   Optimum numbers of nodes for elements: complete polynomials

In this book shape functions are regarded as one of the most important components of the finite-element method. Their convenience in building up piecewise-linear approximations has already been demonstrated in the previous two chapters. However, before looking in detail at the construction of higher-order elements it is instructive to consider briefly the explicit representation of approximating polynomials.

The linear approximation $u = u_i n_i$ used in section 2.1 may, of course, be written in the simpler form

$$u = c_1 + c_2 x + c_3 y \tag{4.1}$$

The expression (4.1) is referred to as a *complete linear polynomial* in two independent variables, since it contains all possible terms of degree one and less. The three independent coefficients c_1, c_2, c_3 can be determined by specifying three values of u, which explains why the three-node triangle is the 'natural' linear element in two dimensions. Similarly, the complete *quadratic* polynomial approximation in two variables is

$$u = c_1 + c_2 x + c_3 y + c_4 x^2 + c_5 xy + c_6 y^2 \tag{4.2}$$

Since there are six independent coefficients c_i, the alternative form $u = u_i n_i$ requires six independent nodal values u_i, the shape functions n_i being quadratic polynomials. More generally, the complete polynomial of degree P in two variables has $(P+1)(P+2)/2$ independent coefficients and requires that number of independent nodal values to specify it uniquely.

A similar sequence of numbers exists for three dimensions. The complete linear polynomial $u = c_1 + c_2 x + c_3 y + c_4 z$ has four independent coefficients, while the complete quadratic polynomial has 10. The number of independent coefficients may be set out in a table

	Linear	Quadratic	Cubic
Two dimensions	3	6	10
Three dimensions	4	10	20

The numbers in this table give the number of nodal values required to define a complete polynomial within an element. Elements which do not have the specified number of nodes give rise to incomplete polynomial approximations – that is, polynomials in which either some of the terms are missing or some of the coefficients c_i are not linearly independent.

The advantage of a complete polynomial approximation is that it is not linked to a particular set of coordinate axes – a function expressed as a complete polynomial in one set of Cartesian axes x, y, z will transform into a complete polynomial of the same degree (with different coefficients) in another set of Cartesian axes x', y', z'. This means that a solution obtained from elements using complete polynomials will not be affected by the choice of coordinate axes used to define the problem.† However, many elements in current use are associated with incomplete polynomial approximations and produce quite acceptable results.

4.2   Simple quadratic and cubic triangles

In the previous section it was shown that a two-dimensional element giving a complete quadratic approximation requires six nodes. This suggests a hexagonal element with straight sides as the natural successor to the linear triangle of the previous two chapters. However, such an element does not ensure continuity in the value of the dependent variable, since each of its boundary segments has only two nodes – not enough to ensure that two connected elements will develop the same quadratic variation along the two sides of their common boundary. To

† The invariance of the scalar products $\mathbf{b}_i \cdot \mathbf{b}_j$ in (2.21) is a good example of this feature of complete polynomials.

Fig. 4.1. A six-node (quadratic) triangle with local coordinates α and β.

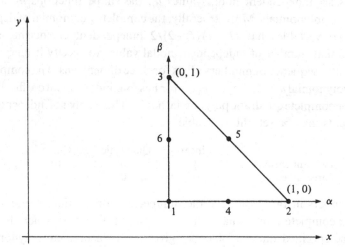

ensure continuity at all points on a straight segment of boundary when the dependent variable varies quadratically within each element requires continuity of value at *three* points on the segment.† This implies that a quadratic element should have three nodes on each straight boundary segment – a condition met by taking a triangle with nodes at the vertices and adding one extra node on each edge. Such a triangle is shown in Fig. 4.1. In this triangle two of the sides are parallel to the x, y axes, and the nodes 4, 5, 6 are positioned at the mid-points of the sides.

At this point it is convenient to introduce local coordinates α, β, as shown in the figure. The quadratic shape functions $n_i(\alpha, \beta)$ are defined in a similar manner to the linear shape functions introduced in chapter 2 – i.e. $n_i(\alpha, \beta)$ is the quadratic polynomial which has value 1 at node i and is zero at the other five nodes. It is easy to verify that the functions

$$\left. \begin{aligned} n_1 &= (1-\alpha-\beta)(1-2\alpha-2\beta) \\ n_2 &= \alpha(2\alpha-1) \\ n_3 &= \beta(2\beta-1) \\ n_4 &= 4\alpha(1-\alpha-\beta) \\ n_5 &= 4\alpha\beta \\ n_6 &= 4\beta(1-\alpha-\beta) \end{aligned} \right\} \tag{4.3}$$

satisfy this definition. The quadratic approximating function is now defined in terms of the local coordinates α, β as $u(\alpha, \beta) = u_i n_i$ if u is a scalar (chapter 2) or $\mathbf{u}(\alpha, \beta) = \mathbf{u}_i n_i$ if $\mathbf{u}$ is a vector (chapter 3). Since the variable u or $\mathbf{u}$ varies quadratically within the element the flow or strain components vary linearly, and indeed this element is often referred to in stress analysis as a *linear strain triangle*.

A cubic triangular element can be developed in a similar way, with nodes at the corners and *two* nodes on each edge, making nine in all. This is one less than the 10 nodes required to produce the complete cubic approximation

$$u = c_1 + c_2\alpha + c_3\beta + c_4\alpha^2 + c_5\alpha\beta + c_6\beta^2 \\ + c_7\alpha^3 + c_8\alpha^2\beta + c_9\alpha\beta^2 + c_{10}\beta^3$$

The deficiency may be rectified by including an internal node 10 at the centre of gravity of the triangle, as shown in Fig. 4.2. The shape function associated with this node has the form $\alpha\beta(1-\alpha-\beta)$. Since this function is zero on all the boundaries of the element it does not affect the boundary continuity conditions. The processing of elements with internal nodes is discussed in section 4.8.

---

† Just as there is only one straight line through two given points, there is only one quadratic function through three.

If the elements of Fig. 4.1 or Fig. 4.2 are used in the construction of a finite-element mesh for a plane region then each element must be paired with an element which has vertices $(0,0)$, $(-1,0)$, $(0,-1)$ in local coordinates. The determination of shape functions for this element is left as an exercise for the reader.

4.3   The four-node square

Although it has the attraction of simplicity, the square element with four nodes has one serious drawback. With a node at each corner, as shown in Fig. 4.3a, only linear variation of u can be allowed on each edge if there is to be continuity between adjacent elements. This means that u must be of the form $c_1 + c_2 \alpha + c_3 \beta + c_4 \alpha \beta$, with four independent parameters corresponding to the four nodes. It is not possible to introduce terms involving α^2 and β^2 – terms necessary to make u a complete quadratic polynomial.

The shape functions for the four-node square are easily seen to be

$$n_1 = (1-\alpha)(1-\beta)/4$$
$$n_2 = (1+\alpha)(1-\beta)/4$$
$$n_3 = (1+\alpha)(1+\beta)/4$$
$$n_4 = (1-\alpha)(1+\beta)/4$$

or

$$n_i = (1+\alpha_i \alpha)(1+\beta_i \beta)/4 \quad (i = 1, ..., 4) \tag{4.4}$$

where α_i, β_i are the coordinates of vertex i. The function n_1 is shown in Fig. 4.3b, the other functions being of similar form.

Fig. 4.2. A 10-node (cubic) triangle.

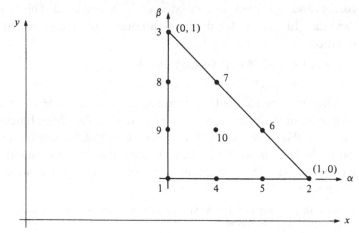

The lack of the α^2 and β^2 terms causes solutions obtained from this element to be influenced by the choice of coordinate axes, as described in section 4.1. In many cases the addition of the $\alpha\beta$ term does not give a great improvement over the linear approximation provided by the three-node triangle. Consider, for example, the use of the square element of Fig. 4.3a to model a section of a beam in pure bending. If the nodes of the element have displacements

$$\mathbf{u}_1 = \begin{bmatrix} \Delta \\ 0 \end{bmatrix}, \quad \mathbf{u}_2 = \begin{bmatrix} -\Delta \\ 0 \end{bmatrix}, \quad \mathbf{u}_3 = \begin{bmatrix} \Delta \\ 0 \end{bmatrix}, \quad \mathbf{u}_4 = \begin{bmatrix} -\Delta \\ 0 \end{bmatrix}$$

then the function $\mathbf{u}$ is given by $\mathbf{u} = \mathbf{u}_i n_i = \begin{bmatrix} \alpha\beta\Delta \\ 0 \end{bmatrix}$. This deformation pattern, shown in Fig. 4.4a, gives the correct linear variation of $\varepsilon_{\alpha\alpha}$ and $\sigma_{\alpha\alpha}$ through the depth of the beam, but that is almost its only resemblance to the exact solution of the pure bending problem, shown in Fig. 4.4b. In the exact solution there are no shear deformations or shear stresses – in the four-node square of Fig. 4.4a there are both. Indeed, the deformation pattern shown in Fig. 4.4a implies shearing tractions on all the edges, with a compensating distribution of body loading to maintain equilibrium. The loading systems associated with the two deformation patterns are shown

Fig. 4.3. (a) A four-node square.
(b) The shape function n_1 for a four-node square.

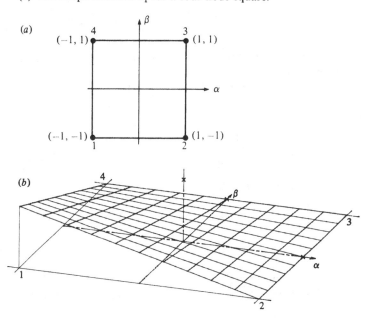

in Figs 4.5a and 4.5b. A modification which has the effect of removing the stiffness due to the spurious shear stresses is described in section 7.1.

4.4   Higher-order square elements

To obtain a more accurate approximation to the displacements associated with plane bending it is necessary to have an element in which straight boundaries can become curved – i.e. an element with quadratic variation along each edge, or at least a pair of opposite edges. Such an

Fig. 4.4. (a) The deformation pattern for a four-node square with nodal displacements corresponding to pure bending.
(b) The correct deformation pattern for pure bending.

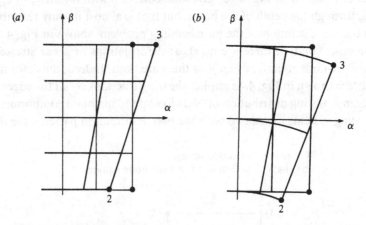

Fig. 4.5. (a) The loading system required to produce the deformation shown in Fig. 4.4a.
(b) The loading system associated with the correct deformation pattern.

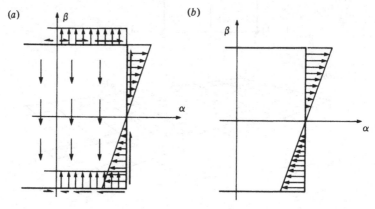

element is the eight-node square shown in Fig. 4.6a. The shape functions n_1 and n_5 are given by

$$n_1 = -(1+\alpha+\beta)(1-\alpha)(1-\beta)/4$$
$$n_5 = (1-\alpha^2)(1-\beta)/2$$

the other functions being derived from these by suitable sign changes.

The eight-node square is one of a class called *serendipity* elements. It

Fig. 4.6. Higher-order square elements:
(*a*) an eight-node serendipity element,
(*b*) a nine-node Lagrangean element,
(*c*) a six-node anisotropic element.

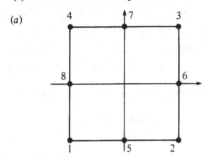

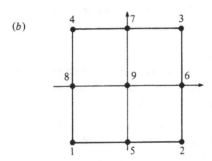

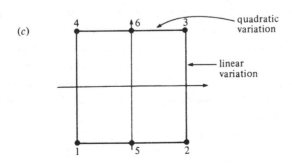

is also possible to construct a nine-node *Lagrangean* square element by including a centre node and using products of simple one-dimensional interpolating polynomials. This element is shown in Fig. 4.6b, the associated shape functions being

$$n_1 = \alpha(1-\alpha)\beta(1-\beta)/4$$
$$n_5 = -(1-\alpha^2)\beta(1-\beta)/2$$
etc.
$$\vdots$$
$$n_9 = (1-\alpha^2)(1-\beta^2)$$

Another possibility is the six-node square shown in Fig. 4.6c. This provides a reasonable model for both 'shearing' and 'bending' deformations. Note that although there are six nodes the displacement approximation is not the complete quadratic provided by the six-node triangle. There is in fact no β^2 term, but the element does allow the displacement pattern $\mathbf{u} = \begin{bmatrix} -\alpha\beta \\ \alpha^2/2 \end{bmatrix}$, which is the pattern associated with pure bending according to the 'plane sections remain plane' assumption. The fact that the polynomial approximation is not a complete quadratic is of little consequence here. It is only sensible to use an element of this sort in a situation where it is known in advance that the general pattern of deformations is of the restricted form provided by the element.

4.5   Three-dimensional elements

The quadratic tetrahedron is a natural extension of the quadratic triangle – the tetrahedron corresponding to the triangle of Fig. 4.1 is shown in Fig. 4.7. There are 10 nodes: the shape functions are easily constructed by adding suitable terms in γ to those given in (4.3) for the quadratic triangle. The 10 nodes provide just enough independent parameters to produce a complete quadratic polynomial, including as independent terms the functions 1, α, β, γ, α^2, β^2, γ^2, $\alpha\beta$, $\beta\gamma$, $\gamma\alpha$. Since there are six nodes on each face (sufficient to define a general quadratic function in a plane), u is continuous in value across each inter-element boundary.

The cubic triangle may also be extended to give the cubic tetrahedron. This has 20 nodes – 4 at the vertices, 12 on the edges at the one-third points and 4 at the mid-points of the faces. This again is just sufficient number to give a complete cubic polynomial, and the presence of 10 nodes on each face ensures continuity in the value of u.

The natural extension of the four-node square is the eight-node cube, with a node at each corner. It is easy to construct the shape functions for this cube from those already given for the square in (4.4), the result being expressible in the form

$$n_i(\alpha, \beta, \gamma) = (1+\alpha_i\alpha)(1+\beta_i\beta)(1+\gamma_i\gamma)/8 \qquad (4.5)$$

where the vertices have coordinates $\alpha_i, \beta_i, \gamma_i = \pm 1, \pm 1, \pm 1$ ($i = 1, ..., 8$). Expansion of this expression shows that the eight-node cube is defective in exactly the same way as the four-node square – the expression in (4.5) gives terms in 1, α, β, γ, $\alpha\beta$, $\beta\gamma$, $\gamma\alpha$, $\alpha\beta\gamma$ but no terms in α^2, β^2, γ^2. Thus the eight-node cube, like the four-node square, cannot reproduce 'bending' deformations.

The three-dimensional extension of the eight-node square is the 20-node cube shown in Fig. 4.8. This element is complete as far as the quadratic terms are concerned, but unlike the cubic tetrahedron (which has the same number of nodes), it does not give all the cubic terms. It is nevertheless a popular element for practical three-dimensional analysis.

In some cases it may be known in advance that a solution is approximately linear (say) in a particular direction. In such cases it may be economical to use elements which are not geometrically isotropic – examples are the 16-node brick and the 15-node wedge shown in Fig. 4.9.

4.6 Two simple tests applicable to higher-order elements

In the previous sections the discussion of higher-order shape functions centred on the need to provide continuity of the dependent variable *on the inter-element boundaries*. This is one of the classical conditions set out in section 1.3 for the convergence of the Ritz process.

Fig. 4.7. A ten-node (quadratic) tetrahedron.

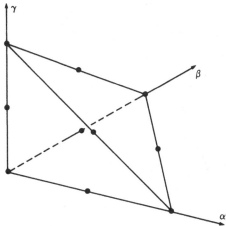

It is equally important for a set of shape functions to generate an approximating function which is satisfactory *within* the associated element.

An approximating function which cannot represent a simple variation of the dependent variable is unlikely to provide a satisfactory approximation when that variation is more complex. It is reasonable to suggest that any higher-order element should reproduce the *exact* solution under conditions of constant flow or stress, since it is evident that the simple linear elements described in chapters 2 and 3 possess this property. This requirement leads

Fig. 4.8. A 20-node cubic element.

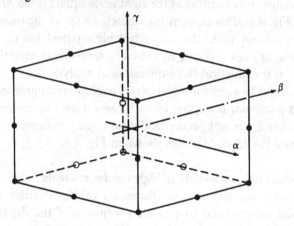

Fig. 4.9. Examples of anisotropic three-dimensional elements.

(a)

(b)

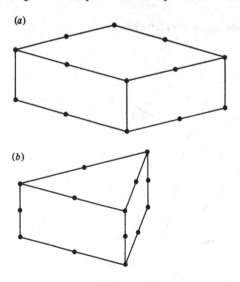

to two simple conditions which can be used to detect erroneous or unsatisfactory sets of shape functions.

Consider an element subject to a linear variation of potential

$$u^L = c_0 + \mathbf{c}_1 \cdot \mathbf{r} \tag{4.6}$$

where $\mathbf{r}$ is the position vector of an arbitrary point (α, β, γ) and c_0, $\mathbf{c}_1$ are constants ($\mathbf{c}_1$ being the uniform potential gradient). If the element has nodes at points $\mathbf{r}_i$, the associated nodal potential values are given by (4.6) as

$$u_i^L = c_0 + \mathbf{c}_1 \cdot \mathbf{r}_i \tag{4.7}$$

Let n_i be the shape functions associated with the element. The approximating polynomial u which takes the values u_i^L at the nodes is

$$u = u_i^L n_i = (c_0 + \mathbf{c}_1 \cdot \mathbf{r}_i) n_i$$

If u is equal to u^L for all values of $\mathbf{r}$ then

$$(c_0 + \mathbf{c}_1 \cdot \mathbf{r}_i) n_i = c_0 + \mathbf{c}_1 \cdot \mathbf{r} \tag{4.8}$$

which implies

$$\Sigma n_i = 1 \tag{4.9a}$$

$$\mathbf{r}_i n_i = \mathbf{r} \tag{4.9b}$$

for any arbitrary point $\mathbf{r}$ within the element. It is easy to check that all the sets of shape functions described in the earlier sections of this chapter satisfy conditions (4.9).

Exactly the same argument applies to elements used for stress analysis. A constant stress field implies a linear variation of displacement, which must be of the form

$$\mathbf{u}^L = \mathbf{c}_0 + \mathbf{C}_1 \mathbf{r} \tag{4.10}$$

where $\mathbf{c}_0$ is a vector and $\mathbf{C}_1$ is a matrix. Conditions (4.9) remain unchanged.

A stress-analysis element whose shape functions satisfy (4.9) will automatically give the correct displacement field (with zero strain) under conditions of rigid-body displacement. A rigid-body displacement field in which the displacements are 'small' (a basic requirement of linear elasticity theory) can be written in the form

$$\mathbf{u}^L = \mathbf{c}_0 + \mathbf{\Omega} \times \mathbf{r} \tag{4.11}$$

where $\mathbf{c}_0$ represents the rigid-body translation and $\mathbf{\Omega}$ the (small) rotation. If equation (4.11) is written in matrix notation it becomes identical to (4.10), with $\mathbf{C}_1$ equal to the anti-symmetric matrix

$$\begin{bmatrix} 0 & -\Omega_\gamma & \Omega_\beta \\ \Omega_\gamma & 0 & -\Omega_\alpha \\ -\Omega_\beta & \Omega_\alpha & 0 \end{bmatrix}$$

4.7   Equivalent nodal sources and loads for higher-order elements

In the physical assembly process of section 2.1 all sources distributed along element boundaries were simply transferred unchanged to the nearest node. In later sections it was suggested that the 'lever rule' was a more rational way of replacing boundary sources or loads. In this section a more general approach is developed which can be used with approximating functions of any order. The analysis will be developed in the language of stress analysis – the equivalent development in terms of thermal or electrical flow is left to the reader.

The phrase 'equivalent systems of forces' runs through a great deal of structural mechanics. Often the word 'equivalent' is used to mean 'statically equivalent', and St. Venant's principle is invoked to justify the substitution of one system of forces for another. However, two statically equivalent systems can have quite different *local* effects. A more precise definition of the equivalence of two sets of forces requires the specification of an allowable set of displacements.

Consider, for example, the replacement of the uniformly distributed load in Fig. 4.10a by the three concentrated forces w_1, w_2, w_3 in Fig. 4.10b. These two load systems are defined to be equivalent if both systems do the same amount of work in any virtual displacement of an appropriately restricted form. A restriction on the form of displacement is necessary, since only

Fig. 4.10. (*a*) A uniform load distributed on an element boundary.
(*b*) Equivalent nodal loads for a boundary with quadratic displacement variation.
(*c*) Equivalent nodal loads for a boundary with cubic displacement variation.

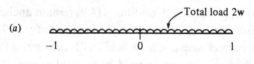

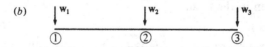

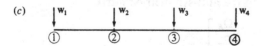

identical load systems do the same amount of work in a completely arbitrary displacement.

If the boundary 1–2–3 in Fig. 4.10*a* is part of a quadratic element the displacement of the two load systems will automatically be restricted to a general quadratic curve, which can be written as a linear combination of the three shape functions

$$n_1 = (\alpha - 1)\alpha/2, \quad n_2 = 1 - \alpha^2, \quad n_3 = (1 + \alpha)\alpha/2$$

shown in Fig. 1.1. Treating each of these functions in turn as a virtual displacement generates the three virtual work equations

$$\mathbf{w}_i = \int_{-1}^{1} n_i \mathbf{w} \, d\alpha \quad (i = 1, 2, 3) \tag{4.12}$$

giving $\mathbf{w}_1 = \mathbf{w}_3 = \mathbf{w}/3$, $\mathbf{w}_2 = 4\mathbf{w}/3$. Equation (4.12) is similar to the equation for $\mathbf{w}_i$ in (3.14), and indeed the analysis presented above is nothing more than a one-dimensional version of the analysis given in section 3.2. If the distributed load is replaced by *four* equally-spaced concentrated loads, as shown in Fig. 4.10*c*, and a cubic displacement variation is assumed, the equivalent system is given by $\mathbf{w}_1 = \mathbf{w}_4 = \mathbf{w}/4$, $\mathbf{w}_2 = \mathbf{w}_3 = 3\mathbf{w}/4$. The verification of this result is left as an exercise for the reader.

The concept of 'statical equivalence' may now be seen as a special case of this more general definition of equivalence. In statical equivalence the only displacement allowed is a rigid-body one, and with such a displacement there are only two independent displacement modes – uniform displacement and uniform rotation. Thus any system of forces can be reduced to a statically equivalent force and couple, but the second system ceases to be 'equivalent' to the first if other modes of displacement are permitted.

Perhaps the most familiar example of this idea comes from elementary bending theory. An engineer replaces the axial stress distribution on the end of a beam by an 'equivalent' axial force and couple, probably without giving the matter much thought. However, the legitimacy of the replacement rests on the assumption that axial displacements are such that 'plane sections remain plane', which is essentially a restriction imposed on the displacements associated with the end loading.

4.8 Assembling the nodal equations

The analysis of chapters 2 and 3 is not greatly altered by the use of higher-order elements. The general nature of the change can be seen by considering the use of the six-node triangle of Fig. 4.1 in place of the corresponding linear triangle in the analysis of section 2.2.

The only changes in the analysis of that section are the use of local coordinates α, β in place of global coordinates x, y, the introduction of piecewise-quadratic nodal shape functions in place of piecewise-linear ones and an increase in the range of the subscripts referencing the nodes. The change of coordinate system makes no difference to the analysis, since the nodal properties of an element are independent of the position of the origin of coordinates. The contribution of a single element to the nodal equations is given by a modified version of (2.20a),

Left-hand side, column j' Right-hand side

$$\text{Add to row } i' \quad \left[D \int_0^1 \int_0^{1-\beta} \mathbf{b}_i \cdot \mathbf{b}_j \, d\alpha \, d\beta \right] u_{j'} \quad \int_0^1 \int_0^{1-\beta} w n_i \, d\alpha \, d\beta$$

(4.13)

in which $d\alpha \, d\beta$ replaces dA, the limits of integration have been put in explicitly and the local node numbers i, j take the values $1, \ldots, 6$. The vectors $\mathbf{b}_i = \nabla n_i$ in (4.13) are linear functions of α and β rather than constants, since the element shape functions n_i given in (4.3) are quadratic.† Equations (2.21) become

$$k_{ij} = D \int_0^1 \int_0^{1-\beta} \mathbf{b}_i \cdot \mathbf{b}_j \, d\alpha \, d\beta \tag{4.14a}$$

$$w_i = \int_0^1 \int_0^{1-\beta} w n_i \, d\alpha \, d\beta \quad (i, j = 1, \ldots, 6) \tag{4.14b}$$

Since the vectors $\mathbf{b}_i$ are linear the integrands in (4.14a) are quadratic functions of α and β. If w is constant the integrands in (4.14b) are also quadratic. These integrals can be evaluated analytically, but it is more common to use a Gauss integration formula. Since a Gauss formula involving four points is exact for a quadratic function in two dimensions, only four evaluations of each integrand are required to give an exact result. There are 36 coefficients k_{ij} to be inserted in the nodal equations, the pattern being an obvious extension of that given in (2.20b). Only 21 separate integral evaluations are required if account is taken of the symmetry property $k_{ij} = k_{ji}$.

If the same element configuration is used for plane stress analysis, equations (4.14) are replaced by analogous equations derived from (3.14)

$$\mathbf{K}_{ij} = \int_0^1 \int_0^{1-\beta} \mathbf{B}_i^t \mathbf{D} \mathbf{B}_j \, d\alpha \, d\beta \tag{4.15a}$$

$$\mathbf{w}_i = \int_0^1 \int_0^{1-\beta} n_i \mathbf{w} \, d\alpha \, d\beta \tag{4.15b}$$

† The operator ∇ can be written as $\begin{bmatrix} \partial/\partial\alpha \\ \partial/\partial\beta \end{bmatrix}$ since the α, β axes are parallel to the x, y axes.

where once again the node numbers i, j take the values $1, \ldots, 6$ and the matrices $\mathbf{B}_i = \square n_i$ are linear functions of α and β. It should be noted that the components of the matrices $\mathbf{B}_i$ are simply the components of the vectors $\mathbf{b}_i$ appearing in (4.14) arranged in a different pattern (see equation (3.12)). Once again, both sets of integrands are quadratic functions of α and β provided that $\mathbf{w}$ is constant.

The overall organisation of the analysis remains exactly the same for other complex elements. A change to three-dimensional elements increases the computation load considerably – a 20-node cube, for example, gives rise to 210 coefficients k_{ij} (or 3×3 matrices $\mathbf{K}_{ij}$) even allowing for the symmetry of the nodal equations.† For this element the calculation of each coefficient or matrix element requires the integration of a quartic function of α and β, since the shape functions n_i have cubic terms which make the corresponding vectors $\mathbf{b}_i$ quadratic. The increase in computational load associated with each element is, of course, offset by the need for fewer elements for a given accuracy of solution.

The equation associated with a particular node p contains terms for all nodes which have an element in common with p. It follows that the use of complex elements increases the number of non-zero coefficients in each equation. If an element (such as a 10-node triangle) has an internal node then the only non-zero coefficients in the equation for that node are those corresponding to nodes belonging to the element.

Internal nodes may be treated in exactly the same way as other nodes during the assembly and solution of the nodal equations. Alternatively the nodal variable associated with an internal node may be eliminated before the element assembly process begins. This procedure generates a modified element with only boundary nodes, which is assembled in the normal way.

It was mentioned in section 2.8 that there is no particular reason why a finite-element solution should be more accurate at the nodes than at other points. Indeed, since the method as developed so far gives discontinuities in $\mathbf{q}$ (or σ) at the nodes, it is obvious that nodal values are not, in general, the most reliable. In practice the most accurate values of u and $\mathbf{q}$ (or $\mathbf{u}$ and σ) in quadratic and higher-order elements are found to be those *at the Gauss points*.‡ Once the nodal equations have been solved, these quantities can be generated quite easily from the nodal variables u_i (or $\mathbf{u}_i$), since the values of the shape functions n_i and their derivatives are available at the Gauss

---

† But see problem 4.5 at the end of this chapter.

‡ In the linear triangular elements of chapters 2 and 3 the centroid of the triangle is, in effect, the single Gauss point required for exact integration of a linear function.

points as a by-product of the integration process. If nodal values are required they can be computed subsequently by interpolation.

4.9   A numerical example

The accuracy of the various finite elements discussed in this and earlier chapters may be assessed by using them to obtain approximate numerical solutions to a problem whose analytical solution is known. A convenient problem in this context is the determination of the deformation of a cantilever of constant rectangular cross-section with a load at the free end, as shown in Fig. 4.11. The stress system is assumed to be two-dimensional.

The classical solution of this problem[†] starts from the assumed stress distribution

$$\sigma_{xx} = Wxy/I \tag{4.16a}$$
$$\sigma_{yy} = 0 \tag{4.16b}$$
$$\tau_{xy} = (W/2I)(d^2/4 - y^2) \tag{4.16c}$$

where $I = bd^3/12$. These equations imply that the load W is applied as a parabolic distribution of shear traction over the free end of the cantilever, while the restraining moment at the root is provided by a longitudinal stress σ_{xx} which varies linearly with y. It is simple to verify that the stress system given in (4.16) satisfies the equilibrium conditions (3.3).

Integrating the strain equations derived from (4.16) gives

$$\mathbf{u} = (W/EI)\begin{bmatrix} x^2y/2 - y^3(2+v)/6 - C_1y + C_2 \\ -vxy^2/2 - x^3/6 + (d^2(1+v)/4 + C_1)x + C_3 \end{bmatrix} \tag{4.17}$$

C_1, C_2 and C_3 being constants of integration. Since the boundary *stresses*

† See, for example, Reference 4, p. 41.

Fig. 4.11. A cantilever carrying a concentrated load at its end.

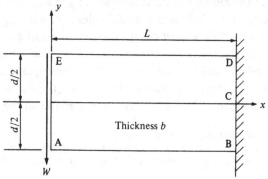

are defined over the whole boundary of the solution region it is not
legitimate to also define boundary *displacements*, except insofar as is
necessary to prevent rigid-body movement of the cantilever. Such movement
is prevented by imposing three displacement constraints, here chosen to
be

$$u_x = u_y = 0 \quad \text{at} \quad x = L, \quad y = 0$$
$$u_x = 0 \quad \text{at} \quad x = L, \quad y = d/2$$

These conditions determine the arbitrary constants in (4.17), giving

$$\mathbf{u} = (W/2EI)\begin{bmatrix} y(x^2 - L^2) + (2+v)\,y(d^2/4 - y^2)/3 \\ (L^3 - x^3)/3 + (L^2 + d^2(4+5v)/12)\,(x-L) - vxy^2 \end{bmatrix}$$
$$(4.18)$$

(Because of symmetry u_x is also zero at $x = L$, $y = -d/2$.) The vertical
displacement of the tip of the cantilever is given by

$$u_y(0,0) = -(WL^3/3EI)(1 + (2+2.5v)\,d^2/4L^2) \qquad (4.19)$$

In carrying out a finite-element analysis of this problem it is important
to replace the distributed tractions over the boundaries AE and BD by
equivalent nodal forces calculated in accordance with the procedure
described in section 4.6. A node on the boundary between B and D, other
than at the point C, must be treated as a normal displacing node, since
in the exact solution (4.18) points on this boundary displace in both the
horizontal and vertical directions. Particular care must be taken over the
calculation of the horizontal components of the nodal forces at nodes on
BD, since these correspond to the linear distribution of σ_{xx} specified in
(4.16a). Omission of these forces produces a significant change in the tip
deflection.

Fig. 4.12 shows the ratio of the approximate tip deflection, as calculated
by various finite-element analyses, to the true tip deflection (4.19), for the
case $L/d = 2$. It will be noticed that in all cases the approximate deflection
is less than the true value, as is to be expected from the analysis developed
in section 3.5. It must be borne in mind that all the terms in the exact
solution (4.18) are present in a complete cubic polynomial, so that in this
example a pair of 10-node cubic triangles, say, would be capable of
generating the exact solution, without any errors other than those due to
numerical round-off. The value of v has been taken as 0.3.

The most striking feature of Fig. 4.12 is the superiority of the six-node
triangle and the eight-node square over the linear triangle and the four-node
square. It is true, of course, that the higher-order elements require more
computational effort for the calculation of the $\mathbf{K}_{ij}$ matrices, but the

advantage of the higher-order elements is still apparent if Fig. 4.12 is re-drawn on the basis of actual computing time taken.

The superiority of the higher-order elements becomes even more marked when stresses are considered. Fig. 4.13 shows plots of the function $\sigma_{xx}I/WLy$ at $y = 0.375d$ for analyses employing three-node triangles, four-node squares and eight-node squares, each analysis being based on the same arrangement of nodal points. The eight-node square is the only one of these three elements with a displacement approximation which allows the longitudinal stress σ_{xx} to take the form given in (4.16a) – a fact which explains the very small error in $\sigma_{xx}I/WLy$ (a maximum of 0.003) for the analysis based on this element.

In this example the prediction of shear stresses provides a severe test for the finite-element approximations. Fig. 4.14 shows the distribution of shear stress on a vertical plane near the centre of the cantilever for the same arrangements of elements. It will be seen that even the assembly of eight-node square elements does not give a good approximation to the correct parabolic distribution. This is because the displacement approximation for the eight-node square has no y^3 term, and it is the y^3 term in the

Fig. 4.12. Numerical results for a loaded cantilever, showing the accuracy obtained using various types and numbers of elements. The pairs of numbers associated with points on the curves give the numbers of elements (or numbers of pairs of elements in the case of triangles) in the x and y directions.

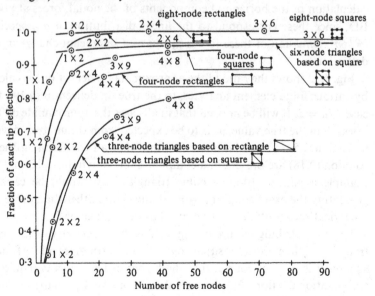

Fig. 4.13. Plots showing the variation of direct stress σ_{xx} with x for various types of element.

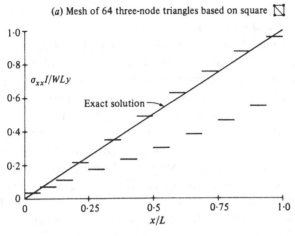

(a) Mesh of 64 three-node triangles based on square

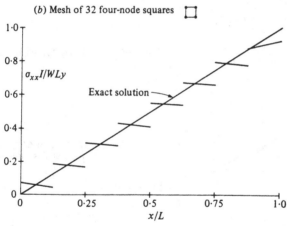

(b) Mesh of 32 four-node squares

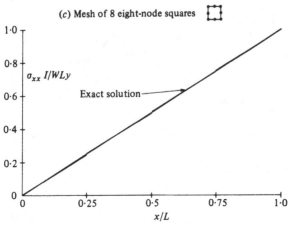

(c) Mesh of 8 eight-node squares

expression for u_x in (4.18) which produces most of the y^2 term in the shear-stress distribution (4.16c).

4.10   The order of convergence

It can be shown that in a finite-element analysis using conforming elements the errors in the approximation u (or $\mathbf{u}$) are of order h^{P+1}, where h is the length of a typical element boundary and P is the degree of the polynomial used in the approximating function. The errors in the first derivatives (i.e. flows or stresses) are therefore of order h^P.

This result makes it possible to extrapolate from the results of two analyses with different mesh sizes. Consider, for example, finite-element analyses based on linear elements. Since $P = 1$, the errors in u vary as h^2. Let $u^{(1)}$ be the solution based on elements of size h and $u^{(2)}$ be the solution based on elements of size $h/2$ (i.e. each triangle in the first analysis is replaced by four triangles in the second). If $\tilde{u}$ is the true solution then

$$\frac{\tilde{u} - u^{(1)}}{\tilde{u} - u^{(2)}} = \frac{O(h^2)}{O((h/2)^2)} \approx 4$$

Fig. 4.14. Plots showing the variation of shear stress τ_{xy} with y for various types of element.

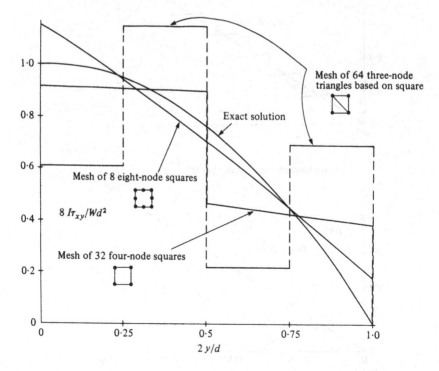

Hence

$$\tilde{u} \approx (4u^{(2)} - u^{(1)})/3 \qquad (4.20)$$

It must be realised that 'of order h^{P+1}' means that the errors are proportional to h^{P+1} as $h \to 0$. For a finite value of h equation (4.20) is only approximate, the accuracy of the approximation depending on the ratio of h to the overall size of the solution region.

As an illustration, the following values are taken from Fig. 4.12, using the results of analyses based on four-node squares,

	tip deflection/exact tip deflection
2×4 pattern ($h = L/4$)	0.878
4×8 pattern ($h = L/8$)	0.934

The four-node square uses an approximating polynomial which is only complete as far as linear terms are concerned, so that $P = 1$. Application of (4.20) gives a deflection prediction which is 0.953 times the true value. The inaccuracy of this prediction is due to the relatively small number of elements used in the analysis.

Note that this procedure only reduces errors due to the finite size of the element – these are normally referred to 'truncation errors'. Application of (4.20) always *increases* the numerical rounding errors.

Problems for chapter 4

4.1 Construct suitable shape functions for the following plane elements:
   (*a*) A six-node triangle with nodes at the vertices $(1, 0)$, $(1, 1)$, $(0, 1)$ and at the mid-points of the sides.
   (*b*) A triangular 'transition element' with nodes at the points $(0, 0)$, $(1, -1)$, $(1, 0)$, $(1, 1)$, giving linear variation on two sides and quadratic variation on the third.
   (*c*) A square transition element with vertices at $(\pm 1, \pm 1)$ and mid-side nodes at $(1, 0)$, $(0, 1)$, giving quadratic variation on two sides and linear variation on the other two.

4.2 A triangular element is defined with nodes at the vertices $(0, 0)$, $(1, 0)$, $(0, 1)$ and nodal variables u, $\partial u/\partial \alpha$, $\partial u/\partial \beta$. (This may be regarded as the limiting case of a nine-node triangle as the nodes on the sides are moved nearer to the corners.) Verify that the shape functions associated with the nine nodal unknowns are
   $$n_1 = (1 - \alpha - \beta)^2 (1 + 2\alpha + 2\beta)$$
   $$n_{\alpha 1} = (1 - \alpha - \beta)^2 \alpha$$
   $$n_2 = \alpha^2 (3 - 2\alpha)$$
   $$n_{\alpha 2} = \alpha^2 (\alpha - 1)$$
   $$n_{\beta 2} = \alpha^2 \beta$$
   etc.,

where $n_{\alpha i}$, $n_{\beta i}$, etc. indicate the shape functions associated with $\partial u/\partial \alpha$ and $\partial u/\partial \beta$ at node i.

4.3 The displacement function associated with the nine variables in problem 4.2 is an incomplete cubic polynomial. It may be completed by the addition of a tenth variable in the form of the displacement at the centroid. What is the shape function associated with this variable? What modifications are required to the shape functions specified in problem 4.2?

4.4 Write down the shape functions for the quadratic tetrahedron shown in Fig. 4.7.

4.5 In a 20-node cubic element used for solving three-dimensional potential problems, how many different numerical values appear in the coefficients k_{ij}?

4.6 Find the equivalent nodal loads for the following distributions:
   (i) a linear distribution $w = 96\alpha$,
   (ii) a parabolic distribution $w = 12(1 - \alpha^2)$,
   assuming the boundary $-1 \leqslant \alpha \leqslant 1$ to be composed of
   (a) three equal segments with linear variation of displacement on each segment,
   (b) two equal segments with quadratic variation on each segment.

4.7 In the analysis of the cantilever shown in Fig. 4.11 the evaluation of the $\mathbf{K}_{ij}$ matrices for eight-node square elements is carried out by a 3×3 Gauss integration procedure. The computer subroutine used for the integration has an error in the Gauss weight associated with the point $(0, 0)$ in element coordinates, this coefficient being 10 times its correct value. What effect would you expect this error to have on the results of analyses in which the cantilever is represented by
   (a) two eight-node squares,
   (b) eight eight-node squares?

4.8 Verify that conditions (4.9) are satisfied by the shape functions for a six-node triangle given in section 4.2.

Solutions to problems

4.1 The elements are shown in Fig. 4.15.
   (a) The shape functions are

$n_1 = (1 - \beta)(1 - 2\beta), \quad n_3 = (1 - \alpha)(1 - 2\alpha)$
$n_2 = (1 - \alpha - \beta)(3 - 2\alpha - 2\beta)$

$n_4 = 4(1-\beta)(\alpha+\beta-1), \quad n_5 = 4(1-\alpha)(\alpha+\beta-1)$
$n_6 = 4(1-\alpha)(1-\beta)$

(*b*) It is simplest to proceed in three stages.

(i) Set up the linear shape functions

$n_1 = 1-\alpha, \quad n_2 = (\alpha-\beta)/2, \quad n_3 = (\alpha+\beta)/2$

(ii) Construct the quadratic shape function $n_4 = \alpha^2 - \beta^2$.

(iii) Subtract sufficient of n_4 from n_2 and n_3 to make them zero at $(1,0)$, giving finally

$n_2 = (\alpha-\beta)/2-(\alpha^2-\beta^2)/2, \quad n_3 = (\alpha+\beta)/2-(\alpha^2-\beta^2)/2$

(*c*) The shape functions are

$n_1 = (1-\alpha)(1-\beta)/4, \quad n_2 = -(1+\alpha)\beta(1-\beta)/4$
$n_4 = -(1+\beta)\alpha(1-\alpha)/4$
$n_3 = (1+\alpha)(1+\beta)(\alpha+\beta-1)/4, \quad n_5 = (1-\beta^2)(1+\alpha)/2$
$n_6 = (1-\alpha^2)(1+\beta)/2$

[Note: It is tempting to write $n_3 = (1+\alpha)(1+\beta)\alpha\beta/4$. However, the resulting set of shape functions does not satisfy the condition $\Sigma n_i = 1$ (equation 4.9*a*) and therefore does not model a uniform potential field correctly.]

4.2 The shape functions in this problem must satisfy *derivative* as well as *value* conditions at the nodes. The required conditions are

$n_1 = 1$ at node 1 and zero at nodes 2 and 3: $\partial n_1/\partial\alpha = \partial n_1/\partial\beta = 0$ at all nodes.
$\partial n_{\alpha 1}/\partial\alpha = 1$ at node 1 and zero at nodes 2 and 3: $n_{\alpha 1} = \partial n_{\alpha 1}/\partial\beta = 0$ at all nodes.
etc.

It is easy to verify that the given functions satisfy these conditions.

4.3 The shape function associated with node 4 is $n_4 = 27\alpha\beta(1-\alpha-\beta)$. Both $\partial n_4/\partial\alpha$ and $\partial n_4/\partial\beta$ are zero at all the corner nodes, so that n_4 satisfies the required value and derivative conditions. The other shape functions must be adjusted by subtracting suitable multiples of n_4 to give each one zero

Fig. 4.15.

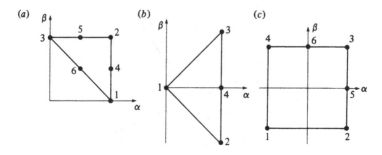

value at the centroid. Putting $\alpha = \beta = \frac{1}{3}$ gives

$n_1 = \frac{7}{27}$, $n_{\alpha 1} = \frac{1}{27}$, $n_{\beta 1} = \frac{1}{27}$, $n_2 = \frac{7}{27}$, $n_{\alpha 2} = -\frac{2}{27}$, $n_{\beta 2} = \frac{1}{27}$, etc.

Hence these multiples of n_4 should be subtracted from the original shape functions to give the final result.

4.4 The element is shown in Fig. 4.16. A simple extension of equations (4.3) gives

$n_1 = (1 - \alpha - \beta - \gamma)(1 - 2\alpha - 2\beta - 2\gamma)$
$n_2 = \alpha(2\alpha - 1)$, $\quad n_3 = \beta(2\beta - 1)$, $\quad n_4 = \gamma(2\gamma - 1)$
$n_5 = 4\alpha(1 - \alpha - \beta - \gamma)$, $\quad n_7 = 4\beta(1 - \alpha - \beta - \gamma)$, $\quad n_8 = 4\gamma(1 - \alpha - \beta - \gamma)$
$n_6 = 4\alpha\beta$, $\quad n_9 = 4\alpha\gamma$, $\quad n_{10} = 4\gamma\beta$

Note that these functions satisfy conditions (4.9).

4.5 There are only two types of node – corner and edge. These are shown in Fig. 4.17. The node pairings are all of the following form:

$C_1 - C_1$ $C_1 - E_1$ $E_1 - E_1$
$C_1 - C_2$ $C_1 - E_2$ $E_1 - E_2$
$C_1 - C_3$ $C_1 - E_4$ $E_1 - E_3$
$C_1 - C_4$ $E_1 - E_4$
                      $E_1 - E_5$

Thus there are 12 essentially different pairings. [Note, however, that if the cube is mapped into a general curvilinear brick element (see chapter 5) all the k_{ij} will, in general, be different, apart from the essential symmetry $k_{ij} = k_{ji}$.]

Fig. 4.16.

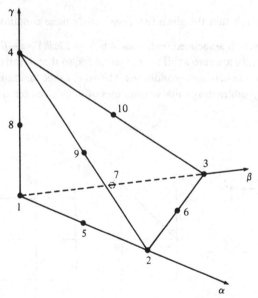

4.6 The loading distributions are shown in Fig. 4.18, and the required shape functions are shown in Fig. 4.19. The equivalent nodal loads are obtained by evaluating (4.12) using appropriate combinations of the given functions. The results are:

(i) (a) $w_3 = -w_1 = \frac{64}{3}$, $w_4 = -w_2 = \frac{224}{9}$

   (b) $w_5 = -w_1 = 16$, $w_4 = -w_2 = 32$, $w_3 = 0$ (from symmetry)

(ii) (a) $w_3 = w_2 = \frac{176}{27}$, $w_4 = w_1 = \frac{40}{27}$

   (b) $w_4 = w_2 = \frac{28}{5}$, $w_5 = w_1 = \frac{1}{5}$, $w_3 = \frac{22}{5}$

Each set of results may be verified by an appropriate statical equilibrium check.

4.7 The integral formula $K_{ij} = \int_A B_i^t D B_j \, dA$ is derived from an integration of the strain energy over the area of the element. Gauss integration replaces this integral by a weighted sum of the values of the integrand at a set of sampling points – the 'Gauss points'. An erroneously large multiplier associated with the central Gauss point will increase the

Fig. 4.17.

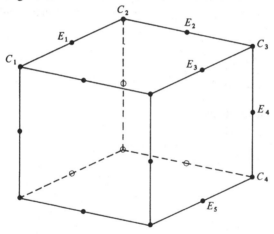

Fig. 4.18. Loading distributions (i) $w = 96\alpha$, (ii) $w = 12(1-\alpha^2)$.

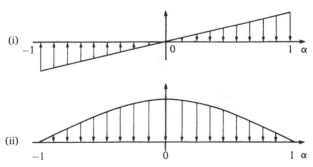

stiffness with respect to any mode which has a non-zero strain-energy density at that point, but will have no effect on the calculated stiffness if the mode is one in which the strain-energy density at the central point is zero.

For an eight-node square this means that the stiffness with respect to extensional deformation and uniform shear deformation will be increased, while stiffness in a symmetrical 'bending' deformation will not be affected. It follows that if a cantilever is represented as two eight-node squares, with centres on the neutral axis, only the shear deformation (which is small) will be reduced. However, if the cantilever is represented by eight eight-node squares the centres of the squares will not be on the neutral axis. Consequently bending of the cantilever will require extensional deformation of each element. The increased extensional stiffness of the elements will result in the deflection of the tip of the cantilever being correspondingly reduced.

4.8 Substitution of the shape functions given in (4.3) into (4.9a) gives $\Sigma n_i = 1$. It is also easy to show that $\alpha_i n_i = (n_4 + n_5)/2 + n_2 = \alpha$, and that $\beta_i n_i = (n_5 + n_6)/2 + n_3 = \beta$. Hence $\mathbf{r}_i n_i = \mathbf{r}$, which is condition (4.9b).

Fig. 4.19. Shape functions:
(a) linear functions

$$n_3 = 3(\alpha + \tfrac{1}{3})/2 \quad -\tfrac{1}{3} \leqslant \alpha \leqslant \tfrac{1}{3}$$
$$ = 3(1 - \alpha)/2 \quad \tfrac{1}{3} \leqslant \alpha$$
$$n_4 = 3(\alpha - \tfrac{1}{3})/2 \quad \tfrac{1}{3} \leqslant \alpha$$

(b) quadratic functions

$$\left.\begin{array}{l} n_3 = (1 - 2\alpha)(1 - \alpha) \\ n_4 = 4(1 - \alpha) \\ n_5 = \alpha(2\alpha - 1) \end{array}\right\} \quad 0 \leqslant \alpha$$

(a)

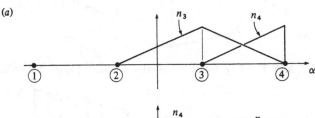

(b)

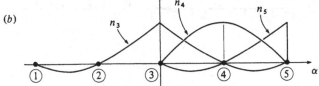

5

Higher-order approximations: (2) Generalising the Element Geometry

The triangles, tetrahedra, squares and cubes of the previous chapter are of limited value as they stand because of their fixed geometry. The first part of the present chapter shows how shape functions, initially developed as approximating polynomials, can also be used as mapping functions to produce elements of more general geometrical form. In particular, it shows how mappings based on the quadratic and cubic shape functions of chapter 4 may be used to generate elements with curved boundaries. The technique may be used in both potential theory and stress analysis.

The generalisation of the element geometry means that Gauss integration of the integrals for the element coefficients k_{ij} or $\mathbf{K}_{ij}$ gives results which are only approximate. At first sight this appears to be an inherent defect of general higher-order elements. However, it is possible to make the errors in the integration procedure balance the errors due to the restricted form of the functional approximation, giving elements which are extremely accurate. This technique, known as reduced integration, is discussed in the second part of the chapter.

In certain types of problem, such as those occurring in the analysis of plates and shells, it is advantageous to use elements which violate the classical Ritz continuity conditions set out in section 1.3. Whether analyses using such elements actually converge can be determined by applying a less restrictive condition known as the patch test. This important condition is described in the concluding section of the chapter.

5.1 Linear mapping of a plane triangle

Fig. 5.1a shows a six-node plane triangle with straight sides, nodes being located at the vertices and at the mid-points of the sides. Imagine that this element is to be used as a quadratic element in the solution of

105

Poisson's equation. The six nodes make the approximation $u = u_i n_i$ a complete quadratic polynomial, the six shape functions n_i being defined in the usual way – i.e. $n_i(x, y)$ is a quadratic polynomial which has value 1 at node i and zero at the other five nodes. The contribution of the element to the nodal equations is given by a modified version of (2.20a),

$$\text{Left-hand side, column } j' \qquad \text{Right-hand side}$$

$$\text{Add to row } i' \quad \left[D \iint_A \mathbf{b}_i \cdot \mathbf{b}_j \, dx \, dy \right] u_{j'} \qquad \iint_A w n_i \, dx \, dy$$

$$(5.1)$$

in which $dx \, dy$ replaces dA, i, j take the values $1, \ldots, 6$ and the vectors $\mathbf{b}_i = \nabla n_i$ are linear functions of x and y rather than constants. Equations (2.21) become

$$k_{ij} = D \iint_A \mathbf{b}_i \cdot \mathbf{b}_j \, dx \, dy, \quad w_i = \iint_A w n_i \, dx \, dy \quad (i, j = 1, \ldots, 6)$$

$$(5.2)$$

The construction of the vector functions $\mathbf{b}_i$ and the subsequent evaluation of the integrals in (5.2) are simplified by regarding the general triangle of Fig. 5.1a as a mapping of the triangle shown in Fig. 5.1b, since the shape functions for the latter triangle have already been developed in chapter 4 (equations (4.3)). Since both triangles in Fig. 5.1 have straight sides all that is needed is a *linear* mapping, and this can be constructed by considering only the *vertex* nodes 1, 2 and 3.

The *linear* shape functions for the three-noded triangle with nodes 1, 2, 3 shown in Fig. 5.1b are

$$n_1^L(\alpha, \beta) = 1 - \alpha - \beta, \quad n_2^L(\alpha, \beta) = \alpha, \quad n_3^L(\alpha, \beta) = \beta \qquad (5.3)$$

Fig. 5.1. A linear mapping of a plane triangle: α and β are parametric coordinates in the x, y plane.

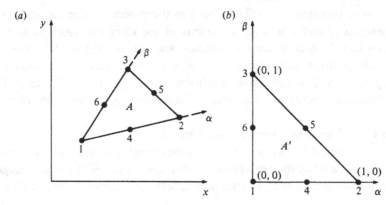

It is clear that the mapping

$$\begin{bmatrix} x \\ y \end{bmatrix} = n_1^L(\alpha, \beta) \begin{bmatrix} x_1 \\ y_1 \end{bmatrix} + n_2^L(\alpha, \beta) \begin{bmatrix} x_2 \\ y_2 \end{bmatrix} + n_3^L(\alpha, \beta) \begin{bmatrix} x_3 \\ y_3 \end{bmatrix} \qquad (5.4a)$$

will map the points 1, 2, 3 in Fig. 5.1b into the corresponding points in Fig. 5.1a. Furthermore, since this mapping is linear it will map any straight line in the α, β plane into a corresponding straight line in the x, y plane. In particular it will map the straight sides of the triangle in Fig. 5.1b into the straight sides of the triangle in Fig. 5.1a and will position nodes 4, 5 and 6 correctly at the mid-points of those sides. Equation (5.4a) may be written more compactly as

$$\begin{bmatrix} x \\ y \end{bmatrix} = n_k^L(\alpha, \beta) \begin{bmatrix} x_k \\ y_k \end{bmatrix} \qquad (k = 1, 2, 3) \qquad (5.4b)$$

From equation (5.4b) it is a matter of formal algebra to obtain the Jacobian matrix of the mapping,

$$\mathbf{J} = \begin{bmatrix} \partial x/\partial \alpha & \partial y/\partial \alpha \\ \partial x/\partial \beta & \partial y/\partial \beta \end{bmatrix} = \begin{bmatrix} x_k \partial n_k^L/\partial \alpha & y_k \partial n_k^L/\partial \alpha \\ x_k \partial n_k^L/\partial \beta & y_k \partial n_k^L/\partial \beta \end{bmatrix} \qquad (5.5)$$

From the definition of the functions n_k^L in (5.3) it follows that $\partial n_k^L/\partial \alpha$, $\partial n_k^L/\partial \beta$ are constant within the element, taking values which are always 0, 1 or -1. This implies that the elements of $\mathbf{J}$ are also constant, as is to be expected in a linear mapping. It is easy to show that $|\mathbf{J}|$ is twice the area A of the triangle shown in Fig. 5.1a.

Having expressed x and y in terms of the coordinates α, β the next step is to express the operator ∇ in a similar way. The required form is†

$$\nabla = \begin{bmatrix} \partial/\partial x \\ \partial/\partial y \end{bmatrix} = \begin{bmatrix} \dfrac{\partial \alpha}{\partial x}\dfrac{\partial}{\partial \alpha} + \dfrac{\partial \beta}{\partial x}\dfrac{\partial}{\partial \beta} \\ \dfrac{\partial \alpha}{\partial y}\dfrac{\partial}{\partial \alpha} + \dfrac{\partial \beta}{\partial y}\dfrac{\partial}{\partial \beta} \end{bmatrix} = \begin{bmatrix} \partial \alpha/\partial x & \partial \beta/\partial x \\ \partial \alpha/\partial y & \partial \beta/\partial y \end{bmatrix}\begin{bmatrix} \partial/\partial \alpha \\ \partial/\partial \beta \end{bmatrix} = \mathbf{J}^{-1}\begin{bmatrix} \partial/\partial \alpha \\ \partial/\partial \beta \end{bmatrix}$$
$$(5.6)$$

The approximating function u is now written in terms of the coordinates α, β as

$$u = n_i(\alpha, \beta)\, u_i \quad (i = 1, \ldots, 6) \qquad (5.7)$$

where the functions n_i are the *quadratic* shape functions given in equations (4.3) for the triangle of Fig. 5.1b. Since the mapping from Fig. 5.1b to Fig. 5.1a is a linear one, the function u is also a quadratic function of x and y. From (5.6)

$$\mathbf{b}_i = \begin{bmatrix} (b_x)_i \\ (b_y)_i \end{bmatrix} = \nabla n_i = \mathbf{J}^{-1}\begin{bmatrix} \partial n_i/\partial \alpha \\ \partial n_i/\partial \beta \end{bmatrix} \qquad (5.8)$$

† Note that in these equations $\partial \alpha/\partial x$, for example, is not the same as $(\partial x/\partial \alpha)^{-1}$.

Since $\mathbf{J}$ is constant over the element and the functions n_i vary quadratically, the components $(b_x)_i$ and $(b_y)_i$ are linear functions of α and β and are easily calculated from (4.3) and (5.8). Note that these components, although expressed as functions of α and β, are still referred to the x and y coordinate axes of Fig. 5.1*a*.

Equations (5.2) may now be written in terms of integrals which have α, β as independent variables.

$$k_{ij} = D \iint_{A'} \mathbf{b}_i \cdot \mathbf{b}_j |\mathbf{J}| \, d\alpha \, d\beta, \quad w_i = \iint_{A'} w n_i |\mathbf{J}| \, d\alpha \, d\beta \qquad (5.9)$$

In each of the left-hand integrals the integrand is a quadratic function of α and β, so that direct analytic integration is still possible. Alternatively a Gauss four-point integration formula can be used – this will also give an exact numerical result because of the form of the integrand. Note that if w is not constant it must be expressed as a function of α and β.

Consider now the use of the triangle in Fig. 5.1*a* as a quadratic displacement triangle for plane stress analysis. The linear mapping procedure is concerned only with the geometry of the element and so remains unchanged. Equations (5.2) are replaced by analogous equations derived from (3.13*b*)

$$\mathbf{K}_{ij} = \iint_A \mathbf{B}_i^t \mathbf{D} \mathbf{B}_j \, dx \, dy, \quad \mathbf{w}_i = \iint_A n_i \mathbf{w} \, dx \, dy \quad (i,j = 1, ..., 6)$$

$$(5.10)$$

where the matrices $\mathbf{B}_i = \square n_i$ are defined in (3.12) as

$$\mathbf{B}_i = \begin{bmatrix} (b_x)_i & 0 \\ 0 & (b_y)_i \\ (b_y)_i & (b_x)_i \end{bmatrix} \qquad (5.11)$$

the quantities $(b_x)_i$, $(b_y)_i$ being the linear functions of α and β given in (5.8). Changing the independent variables in (5.10) from x, y to α, β gives

$$\mathbf{K}_{ij} = \iint_{A'} \mathbf{B}_i^t \mathbf{D} \mathbf{B}_j |\mathbf{J}| \, d\alpha \, d\beta, \quad \mathbf{w}_i = \iint_{A'} n_i \mathbf{w} |\mathbf{J}| \, d\alpha \, d\beta \qquad (5.12)$$

Once again, in each of the integrals on the left-hand side the integrand is a quadratic function of α and β and may be computed exactly either by direct integration or by use of a Gauss four-point formula. As indicated in section 1.1, this formula only requires the evaluation of the integrand at four specified points in the triangle. If $\mathbf{w}$ is not constant then it must be expressed as a function of α and β before the integrations are carried out. The programming of the numerical evaluation of the integrals in (5.9) and (5.12) is discussed in section 8.2.

Note that the variables α, β are only introduced to simplify the integrations. The nodal displacements $\mathbf{u}_j$ and the nodal loads $\mathbf{w}_i$, like the vectors $\mathbf{b}_i$ in (5.8), are still defined by their components in the global x, y coordinate system. Note also that throughout this analysis a clear distinction has been maintained between the *linear* shape functions n_k^L (based on the three vertex nodes) used in the linear mapping (5.4) and the *quadratic* shape functions n_i (based on all six nodes) used in the approximation for u in (5.7). This type of element, in which the mapping uses fewer nodes than the functional approximation, is referred to as a *sub-parametric* element.

5.2 Quadratic mapping of a plane triangle

Consider now an element in the form of a six-node curvilinear triangle, as shown in Fig. 5.2a. To generate this element from the standard triangle of Fig. 5.2b clearly requires a non-linear mapping. The most important requirement of such a mapping is that it should transform the positions of the six nodes correctly. This requirement is satisfied by the mapping

$$\begin{bmatrix} x \\ y \end{bmatrix} = n_k(\alpha, \beta) \begin{bmatrix} x_k \\ y_k \end{bmatrix} \quad (k = 1, ..., 6) \tag{5.13}$$

where the functions n_k are now the *quadratic* shape functions, defined by equations (4.3), which were used in the previous section to set up the approximation u (i.e. equation (5.7)).

The mapping (5.13) converts each straight boundary of the triangle in Fig. 5.2b into part of a conic. For example, the straight boundary

Fig. 5.2. A quadratic mapping of a plane triangle: x and y are quadratic functions of the parametric coordinates α and β.

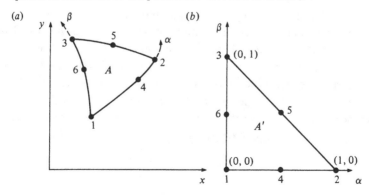

connecting nodes 1, 4 and 2 maps into the curve

$$\begin{bmatrix} x \\ y \end{bmatrix} = n_1(\alpha,0)\begin{bmatrix} x_1 \\ y_2 \end{bmatrix} + n_4(\alpha,0)\begin{bmatrix} x_4 \\ y_4 \end{bmatrix} + n_2(\alpha,0)\begin{bmatrix} x_2 \\ y_2 \end{bmatrix}$$

Substituting for n_1, n_4 and n_2 from (4.3) gives

$$\begin{bmatrix} x \\ y \end{bmatrix} = (1-\alpha)(1-2\alpha)\begin{bmatrix} x_1 \\ y_1 \end{bmatrix} + 4\alpha(1-\alpha)\begin{bmatrix} x_4 \\ y_4 \end{bmatrix} + \alpha(2\alpha-1)\begin{bmatrix} x_2 \\ y_2 \end{bmatrix} \quad (5.14)$$

which is a conic with α as parameter. If two curvilinear triangles have three nodes in common then both the boundary segments containing these nodes will be given by an equation such as (5.14). Thus the use of the mapping (5.13) ensures that the two triangles fit together exactly on their common boundary, even when the boundary is curved.

The analysis of the previous section may now be repeated, with precisely the same quadratic approximation (5.7) for u. The use of the quadratic mapping (5.13) generates Jacobian coefficients $\partial x/\partial\alpha$, etc. which are linear functions of α and β, which makes $|\mathbf{J}|$ a quadratic function of these variables. It follows that the coefficients $(b_x)_i$ and $(b_y)_i$ which appear in (5.8) and (5.11) are each the quotient of two quadratic functions of α and β. Although it is now impracticable to evaluate the integrals in (5.9) and (5.12) analytically there are no particular problems in evaluating them numerically using, say, a Gauss four-point formula. Note that since the integrands are no longer polynomials the result of such numerical integration will not be exact.

The six-node triangle gives continuity of u or $\mathbf{u}$ on the inter-element boundaries, even when these boundaries are curved. For since exactly the same shape functions are used for the mapping as for the functional approximation u, the argument that *boundaries* on adjacent triangles always fit together exactly holds also for *potentials* or *displacements* on those boundaries.

Elements in which the same shape functions are used for the mapping as for the functional approximation are termed *iso-parametric* elements. Although they are one of the most popular types of finite element, it is clear that they are computationally inefficient in situations where their use implies more complex mappings than are really necessary. It is also possible to develop super-parametric elements, in which the order of the shape functions used for the mapping is higher than the order of those used for the functional approximation.

5.3 Sub-parametric and iso-parametric quadrilaterals

The ideas presented in the two previous sections may be used to derive general quadrilateral elements from the square elements of Fig. 4.3 and Fig. 4.6. For example, the shape functions $n_1, \ldots, n_4$ given in equation (4.4) may be used to map the square in Fig. 5.3b into the quadrilateral in Fig. 5.3a, the mapping being given by

$$\begin{bmatrix} x \\ y \end{bmatrix} = n_k(\alpha, \beta) \begin{bmatrix} x_k \\ y_k \end{bmatrix} \quad (k = 1, \ldots, 4) \tag{5.15}$$

Note that this is not a linear mapping unless the quadrilateral is a parallelogram, as it includes terms in $\alpha\beta$. However, all straight lines parallel to the α, β axes, including the boundaries of the square, are mapped into straight lines in the x, y plane.

This mapping may be linked with an approximation u based on the four-node square of Fig. 4.3 to give a four-node iso-parametric quadrilateral. Alternatively the four-node mapping may be applied to any of the elements shown in Fig. 4.6, using the appropriate higher-order approximation. In each case the element characteristics are given by equations very similar to (5.9) or (5.12), the only changes being the range of the subscripts i and j, the region of integration in the α, β plane and the form of the shape functions used in the approximation. The Jacobian matrix $\mathbf{J}$ is not, in general, constant, so that numerical evaluation of the integrals in (5.9) or (5.12) will not be exact.

The four-node mapping (5.15) is sufficient for any straight-sided quadrilateral. If a quadrilateral has one or more curved edges the mapping must include additional nodes on those edges, as in the case of the triangle. The full eight-node iso-parametric quadrilateral is available in most finite-

Fig. 5.3. A mapping of a square into a straight-sided quadrilateral.

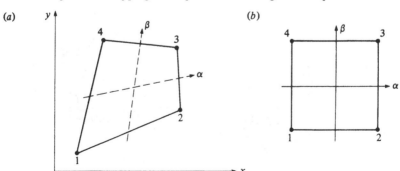

element computer packages, although one can argue that it gives the user more generality in defining meshes than is really necessary. Four curved boundaries are rarely needed in dividing up the solution region.

The extension of the ideas to three dimensions increases the computational load, the matrix $\mathbf{J}$ becoming a 3×3 matrix and the quantities $\mathbf{b}_i$ and $\mathbf{B}_i$ being correspondingly extended. However, the general form of equations (5.9) and (5.12) remains unchanged. The curvilinear iso-parametric '20-node brick' is a popular element in three-dimensional potential flow and stress analysis.

5.4   Ill-conditioning

The curvilinear iso-parametric element is a very versatile tool in finite-element analysis. It allows the finite-element mesh to be graded smoothly from large elements where the dependent variable varies slowly to smaller elements in regions of rapid variation. However, it is advisable to avoid using element shapes which imply a great deal of distortion in the mapping from the basic element shapes of chapter 4. Long, thin needle-like elements, and other elements in which $\mathbf{J}$ varies a great deal, will produce ill-conditioned nodal equations, whether reduced integration is used or not. In extreme cases of gross distortion the mapping may actually have singular points at which $|\mathbf{J}|$ is zero, as shown in Fig. 5.4. Such elements should be broken up into several smaller, less distorted elements.

5.5   Reduced integration

As mentioned earlier, the non-linear nature of the mapping in Fig. 5.2 or Fig. 5.3 means that the integrands in equations (5.9) and (5.12) are

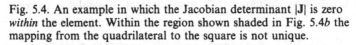

Fig. 5.4. An example in which the Jacobian determinant $|\mathbf{J}|$ is zero *within* the element. Within the region shown shaded in Fig. 5.4*b* the mapping from the quadrilateral to the square is not unique.

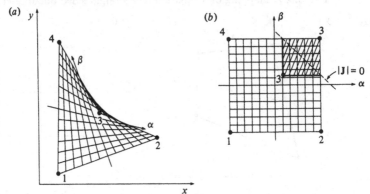

no longer polynomials. Gauss integration formulae do not, therefore, give exact results, however many Gauss points are used. This fact is not really a defect of the analysis, since the restriction on the form of u already introduces a degree of approximation. Indeed, it has been found that even when exact Gauss integration is possible it is beneficial to use an integration formula with *fewer* points than are suggested by the form of u.

Consider, for example, the eight-node square element shown in Fig. 5.4a. The approximation u associated with this element has some (though not all) of the cubic polynomial terms, so that the quantities $\mathbf{b}_i$ and $\mathbf{B}_i$ have components which are quadratic in α and β. If the element is used in its original square form without any mapping then the independent variables α and β are identical with x and y, so that $|\mathbf{J}| = 1$. This means that the integrands in (5.9) and (5.12) are quartic polynomials, requiring a 3×3 (nine-point) Gauss formula for exact evaluation of the integrals.

The choice of an approximation $u = u_i n_i$ for the unknown function u inevitably means that the value of the functional $T(u)$ calculated by the Ritz process is too large – as stated earlier, the true solution $\tilde{u}$ is one which satisfies a minimum energy condition, so that $T(u) \geqslant T(\tilde{u})$. In the language of stress analysis, the eight-node square element, like any other conforming element with a specified displacement field, is too 'stiff'.

This excessive stiffness may be reduced by using a lower-order integration formula – say a 2×2 (four-point) formula, as shown in Fig. 5.5a, in place of the theoretically exact nine-point one. This is an example of the procedure known as *reduced integration*. It can be used with any type of

Fig. 5.5. (a) An eight-node square showing Gauss integration points:
o 3×3 (exact) integration,
x 2×2 (reduced) integration.
(b) A four-node square showing Gauss integration points:
o 2×2 (exact) integration,
x 1×1 (reduced) integration.

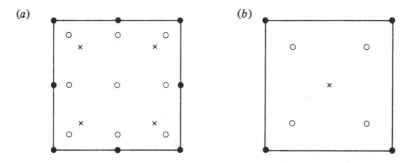

higher-order element, including one involving a non-linear mapping. Although there is usually an improvement in numerical accuracy, there is the accompanying disadvantage that the bounding quality of the approximation is lost – one cannot say whether $T(u)$ will be greater or less than $T(\tilde{u})$, or whether the calculated displacement under a load will be greater or less than the true displacement.

It has already been stated that in the case of an isolated element the coefficients k_{ij} or the matrices $\mathbf{K}_{ij}$ form a singular matrix. To be more specific, in a potential problem a plane element with n nodes generates an $n \times n$ matrix of coefficients k_{ij}: this matrix has rank† $n-1$, since the equivalent nodal flows are not altered when the potential of each node is increased by the same arbitrary amount. Similarly, in a plane stress element the $2n \times 2n$ matrix formed by the matrices $\mathbf{K}_{ij}$ has rank $2n-3$, since the element can be given an arbitrary rigid-body displacement. Reduced integration often introduces an additional degree of singularity into the matrix of element coefficients, as may be seen by considering an approximate solution of the equations of plane stress within the single four-node square element shown in Fig. 5.5b.

This element has eight nodal degrees of freedom. Derivation of the nodal equations starts from the virtual work equations (3.7)

$$\iint_A (\boldsymbol{\varepsilon}^*)^t \, \boldsymbol{\sigma} \, d\alpha \, d\beta = \iint_A (\mathbf{u}^*)^t \, \mathbf{w} \, d\alpha \, d\beta \qquad (5.16)$$

As described in section 3.2, the virtual strain field $\boldsymbol{\varepsilon}^*$ is made equal to $\Box n_i \mathbf{u}_i^*$, giving five linearly independent strain parameters (the eight degrees of freedom associated with $\mathbf{u}_i^*$ minus the three rigid-body degrees of freedom). Hence the $\mathbf{K}_{ij}$ matrices generated from (5.16) form a matrix of rank 5. However, if the integrals in (5.16) are replaced by Gauss approximations based on a single central point, that equation becomes

$$(\boldsymbol{\varepsilon}^*(0,0))^t \, \boldsymbol{\sigma}(0,0) = (\mathbf{u}^*(0,0))^t \, \mathbf{w}(0,0) \qquad (5.17)$$

The quantity $\boldsymbol{\varepsilon}^*(0,0)$ in (5.17) has only three independent parameters, so that the matrices $\mathbf{K}_{ij}$ generated from (5.17) form a matrix of rank 3. The two additional degrees of singularity are associated with the fact that reduced integration gives the element no stiffness with respect to any deformation mode which implies zero strain at the origin – a mode such as that illustrated in Fig. 4.4a.

A similar argument applies if the same element is used to solve a potential problem, in which the dependent variable u is a scalar. There are

---

† The rank of a matrix is simply the number of linearly independent rows. If in a system of N linear equations the coefficient matrix is of rank $M < N$, then $N-M$ of the unknowns can be assigned arbitrary values.

now four nodal degrees of freedom. The virtual work equation (5.16) is replaced by the equation

$$-\iint_A (\mathbf{e}*)^t \mathbf{q} \, d\alpha \, d\beta = \iint_A u^* w \, d\alpha \, d\beta \tag{5.18}$$

In the context of electrical conduction this equation can be described as one of 'virtual power'.† If $\mathbf{e}*$ is made equal to $\nabla n_i u_i^*$ there are three linearly independent parameters associated with (5.18) (the four degrees of freedom associated with the nodal values u_i^* minus a single arbitrary reference potential). The coefficients k_{ij} derived from (5.18) therefore form a matrix of rank 3. However, if the integrals in (5.18) are replaced by their Gauss one-point approximations that equation becomes

$$-(\mathbf{e}*(0,0))^t \mathbf{q}(0,0) = u^*(0,0) \, w(0,0) \tag{5.19}$$

The quantity $\mathbf{e}*(0,0)$ provides only two linearly independent parameters, so that the coefficients k_{ij} derived from (5.19) form a matrix of rank 2.

The singularity of the equations associated with a single element may disappear when a number of elements are joined together, since the assembly process necessarily reduces the number of independent nodal degrees of freedom. Although it is not difficult to check whether a particular application of reduced integration gives a singular set of nodal equations, a mere comparison of the total number of Gauss points against the total number of nodal degrees of freedom may be misleading. The situation is analogous to that of a pin-jointed frame, where an adequate number of bars can form a structure which is redundant in some places and a mechanism in others.

Sometimes the additional degrees of singularity provided by reduced integration can be used to good effect. An example occurs in the development of a deep-beam element in section 7.1. In three-dimensional stress analysis reduced integration has proved useful in cases where the bulk modulus of the material is much greater than the shear modulus. The 'displacement' approach to stress analysis set out in chapter 3 cannot be applied to three-dimensional problems involving incompressible material (i.e. material with Poisson's ratio $v = 0.5$), since coefficients in the matrix $\mathbf{D}$ defined by equation (3.16a) become infinite. However, in cases where v is slightly less than 0.5 reduced integration often gives considerable improvement in accuracy.

---

† Equation (5.18) is mathematically equivalent to equation (2.17) in section 2.2 and could have been used as the starting point for the derivation of the coefficients k_{ij} in that section.

5.6   Relaxing the continuity conditions between elements – the patch test

All the elements developed so far in this book have been *conforming* elements, giving continuity of the dependent variable on the inter-element boundaries. They have also satisfied the conditions (4.9) for the *exact* representation of states of constant flow or stress. It is possible to construct non-conforming elements which do not satisfy conditions (4.9) but nevertheless produce convergence to the true solution as the mesh size is reduced. Such elements are important in many areas of application of the finite-element method.

For such elements it has been shown that if the *average* value of the flow or stress under conditions of uniform potential gradient or strain is correct then the corresponding analysis of a general (non-uniform) distribution will converge to the true solution. This condition is both *necessary* and *sufficient*. The requirement is normally imposed on a group or 'patch' of elements and is known as the *patch test*.

As an illustration of the idea, imagine that Fig. 5.6a represents part of a group of identical triangles in a conducting region where the potential gradient is constant. Let the potential difference between the two lines of nodes be Δu and let the calculated average flow density be $\bar{q}$. In this problem it is meaningless to say that the elements are 'large' or 'small', since there is no ratio of dimensions available to give a measure of size. The replacement of each triangle by four smaller triangles, as shown in Fig. 5.6b, will cause exactly the same flow pattern to be produced on a smaller scale, and will have no effect on the value of $\bar{q}$. Thus if $\bar{q}$ is correct for the mesh of Fig. 5.6a the element will give convergence to the true solution in the general, non-uniform case. The same conclusion holds if Fig. 5.6a represents a portion of a region under conditions of uniform

Fig. 5.6. Flow under conditions of uniform potential gradient.

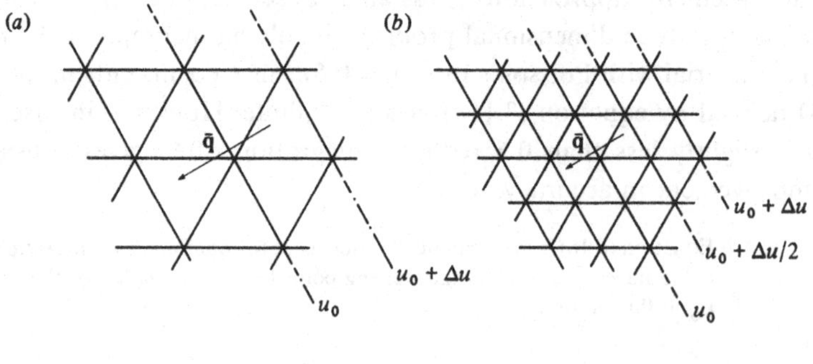

stress – the average strain is not altered by replacing the elements by the smaller elements of Fig. 5.6b.

A non-conforming element which passes the patch test may well be more accurate than a conforming element with the same number of nodes. This is because the discontinuities on the inter-element boundaries tend to balance the inherent excessive 'stiffness' caused by the restricted form of the approximating function. The effect is somewhat similar to that of reduced integration. As with reduced integration, the possible gain in accuracy is accompanied by an uncertainty about the sign of the error.

Problems for chapter 5

5.1 Calculate the coefficients $k_{ij} = DA\mathbf{b}_i \cdot \mathbf{b}_j$ for a general linear triangle by using a linear mapping from a three-node triangle with vertices $(0, 0)$, $(1, 0)$, $(0, 1)$, and a linear approximating function within this triangle. Verify that the results obtained by this procedure agree with those derived in chapter 2 (see solution to problem 2.1).

5.2 If $n_k(\alpha, \beta)$ $(k = 1, \ldots, 4)$ are the shape functions given in equation (4.4) for a four-node square, show that the Jacobian $|\mathbf{J}|$ of the mapping $\begin{bmatrix} x \\ y \end{bmatrix} = n_k(\alpha, \beta) \begin{bmatrix} x_k \\ y_k \end{bmatrix}$ is a linear function of α and β. What is the significance of the constant term in this linear function?

5.3 Determine whether the following examples of reduced integration produce singular nodal equations when used for problems of (a) steady-state heat conduction, and (b) elastic stress analysis.
(i) The block of four-node square elements shown in Fig. 5.7, using a single Gauss point for each square.

Fig. 5.7.

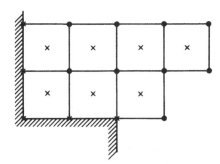

(ii) Two eight-node square elements with a common edge, using four Gauss points for each square.

(iii) A single 20-node cubic element, using eight Gauss points.

In cases (ii) and (iii) assume that the minimum number of external constraints are provided.

5.4 What statements can be made about the mapping

$$\begin{bmatrix} x \\ y \end{bmatrix} = n_k(\alpha, \beta) \begin{bmatrix} x_k \\ y_k \end{bmatrix}$$

if the shape functions $n_k(\alpha, \beta)$ satisfy conditions (4.9), i.e.

$$\Sigma n_k = 1$$

$$n_k(\alpha, \beta) \begin{bmatrix} \alpha_k \\ \beta_k \end{bmatrix} = \begin{bmatrix} \alpha \\ \beta \end{bmatrix}$$

Solutions to problems

5.1 The shape functions for the triangle shown in Fig. 5.8, are $n_1 = 1 - \alpha - \beta$, $n_2 = \alpha, n_3 = \beta$. If $u = u_k n_k$ and $\mathbf{J} = \begin{bmatrix} x_k \partial n_k/\partial \alpha & y_k \partial n_k/\partial \alpha \\ x_k \partial n_k/\partial \beta & y_k \partial n_k/\partial \beta \end{bmatrix}$ equation (5.5)) then

$$\mathbf{J} = \begin{bmatrix} -x_1 + x_2 & -y_1 + y_2 \\ -x_1 + x_3 & -y_1 + y_3 \end{bmatrix}, \quad \mathbf{J}^{-1} = (1/2A) \begin{bmatrix} y_3 - y_1 & y_1 - y_2 \\ x_1 - x_3 & x_2 - x_1 \end{bmatrix}$$

From equation (5.6) $\nabla = \mathbf{J}^{-1} \begin{bmatrix} \partial/\partial \alpha \\ \partial/\partial \beta \end{bmatrix}$ so that

$$\mathbf{b}_1 = \nabla n_1 = \mathbf{J}^{-1} \begin{bmatrix} -1 \\ -1 \end{bmatrix} = (1/2A) \begin{bmatrix} y_2 - y_3 \\ x_3 - x_2 \end{bmatrix}$$

with similar results for $\mathbf{b}_2$ and $\mathbf{b}_3$. These results agree with those obtained in the solution to problem 1, chapter 2. It follows that

$$k_{ij} = D \iint_{A'} \mathbf{b}_i \cdot \mathbf{b}_j |\mathbf{J}| \, d\alpha \, d\beta = D \mathbf{b}_i \cdot \mathbf{b}_j 2AA' = DA \mathbf{b}_i \cdot \mathbf{b}_j$$

which agrees with the expression (2.21) obtained in section 2.2.

Fig. 5.8.

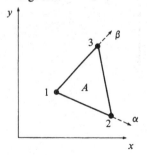
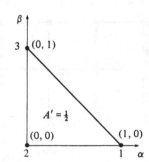

5.2 If $n_i(\alpha, \beta) = (1 + \alpha_i \alpha)(1 + \beta_i \beta)/4$ and $\begin{bmatrix} x \\ y \end{bmatrix} = n_k \begin{bmatrix} x_k \\ y_k \end{bmatrix}$ then from equation (5.5)

$$J = \left(\frac{1}{4}\right) \begin{bmatrix} x_k \alpha_k (1 + \beta_k \beta) & y_k \alpha_k (1 + \beta_k \beta) \\ x_k \beta_k (1 + \alpha_k \alpha) & y_k \beta_k (1 + \alpha_k \alpha) \end{bmatrix}$$

giving

$$|J| = (\tfrac{1}{16}) [x_i \alpha_i (1 + \beta_i \beta) y_j \beta_j (1 + \alpha_j \alpha) - x_i \beta_i (1 + \alpha_i \alpha) y_j \alpha_j (1 + \beta_j \beta)]$$

(Note the use of the separate dummy subscripts i and j to represent the independent summations in the components of J.) The term in $\alpha \beta$ is

$(\tfrac{1}{16}) [x_i \alpha_i \beta_i y_j \beta_j \alpha_j - x_i \beta_i \alpha_i y_j \alpha_j \beta_j] = 0$

Hence $|J|$ is linear in α and β. The full expansion is

$$|J| = (\tfrac{1}{16}) [(x_i \alpha_i y_j \beta_j - x_i \beta_i y_j \alpha_j) + (x_i \alpha_i y_j \beta_j \alpha_j - x_i \beta_i \alpha_i y_j \alpha_j) \alpha \\ + (x_i \alpha_i \beta_i y_j \beta_j - x_i \beta_i y_j \alpha_j \beta_j) \beta]$$

The first term of this expression reduces to

$(\tfrac{1}{8}) [(x_1 y_2 - x_2 y_1) + (x_2 y_3 - x_3 y_2) + (x_3 y_4 - x_4 y_3) \\ + (x_4 y_1 - x_1 y_4)] = $ Area of quadrilateral/4

This is equal to the average magnification of the mapping, i.e. the magnification at the origin $\alpha = 0$, $\beta = 0$.

5.3 (i) (a) If the given mesh is used in a heat-conduction problem there are nine nodal degrees of freedom. Each of the seven integration points provides two independent items of information about the properties of the associated element. Since $2 \times 7 = 14 > 9$ it would appear that the nodal equations are non-singular. It is necessary, however, to check that all the nodes are sufficiently connected for there to be no local singularities. The most likely place for a singularity to occur is in the element on the extreme right of Fig. 5.7. Here there are two nodal degrees of freedom and one integration point linking the associated nodes to the rest of the mesh. This is just sufficient for the equations to be non-singular.

(b) If the mesh is used in a plane stress analysis there are 18 nodal degrees of freedom and three items of information associated with each of the seven integration points. Since $3 \times 7 > 18$ it appears that the nodal equations are non-singular. However, in the element on the extreme right there is only one integration point, giving only three independent items of information about four nodal degrees of freedom. Thus a local mechanism exists and the nodal equations are singular.

(ii) (a) In a heat-conduction problem there are $13 - 1 = 12$ degrees of freedom (one external constraint is required) and 2×8 independent items of information from the Gauss points. The equations are therefore non-singular.

(b) For plane stress analysis the corresponding figures are $13 \times 2 - 3 = 23$ degrees of freedom and 3×8 independent items of information from the Gauss points. Once again the equations are non-singular.

(iii) (*a*) In a three-dimensional heat conduction problem there are twenty nodal degrees of freedom and one external constraint. Each of the eight Gauss points provides three independent items of information. Since $20 - 1 < 3 \times 8$ the nodal equations are non-singular.

(*b*) In a three-dimensional stress-analysis problem there are sixty nodal degrees of freedom and six external constraints. Each of the eight Gauss points provides six independent items of information. Since $60 - 6 > 6 \times 8$ the nodal equations are singular.

5.4 The conditions given are necessary and sufficient for the mapping to be linear *everywhere* if the mapping of the *nodes* is linear. For if

$$\begin{bmatrix} x_k \\ y_k \end{bmatrix} = \mathbf{c}_0 + \mathbf{C}_1 \begin{bmatrix} \alpha_k \\ \beta_k \end{bmatrix}$$

where $\mathbf{c}_0$ and $\mathbf{C}_1$ are constant (a linear mapping of the nodes) and

$$\begin{bmatrix} x \\ y \end{bmatrix} = n_k(\alpha, \beta) \begin{bmatrix} x_k \\ y_k \end{bmatrix}$$

it follows that

$$\begin{bmatrix} x \\ y \end{bmatrix} = \mathbf{c}_0 \Sigma n_k(\alpha, \beta) + \mathbf{C}_1 n_k(\alpha, \beta) \begin{bmatrix} \alpha_k \\ \beta_k \end{bmatrix} \tag{5.20}$$

If $\Sigma n_k(\alpha, \beta) = 1$ and $n_k(\alpha, \beta) \begin{bmatrix} \alpha_k \\ \beta_k \end{bmatrix} = \begin{bmatrix} \alpha \\ \beta \end{bmatrix}$ (conditions (4.9)), equation (5.20) becomes

$$\begin{bmatrix} x \\ y \end{bmatrix} = \mathbf{c}_0 + \mathbf{C}_1 \begin{bmatrix} \alpha \\ \beta \end{bmatrix}$$

Thus a linear mapping of the nodes implies a linear mapping of a general point α, β.

6

Axial symmetry and harmonic analysis

Many three-dimensional problems in potential theory and elastic stress analysis have some degree of radial symmetry and are best analysed using cylindrical polar coordinates r, z, ϕ. Such problems fall into two classes,

(*a*) Those in which the solution region, the boundary conditions and the external source or loading distribution are all symmetric about the z axis, so that the solution also has complete radial symmetry, with r and z as the two independent variables. The solution of such problems is very similar to the solution of the analogous problems in plane potential theory or plane stress.

(*b*) Those in which the solution region and boundary conditions have radial symmetry, but the external source or loading distribution has not. In such problems the source or loading distribution may be expressed as a sum of harmonics in ϕ, i.e. as a Fourier series. The solution for each harmonic is again a problem in two independent variables.

The present chapter extends the analysis of earlier chapters to cover both these cases.

The introduction of a curvilinear coordinate system produces an important change in the relationship between the gradient and divergence operators, and hence in the meaning of the symbol ∇^2. In a rectangular Cartesian coordinate system the divergence operator is simply the transpose of the gradient operator ∇ (see section 1.1), and the symbol ∇^2 is a natural representation of the scalar product $\nabla^t \nabla$. In cylindrical polar coordinates, however, the two operators take the form

$$\nabla = \text{grad} = \begin{bmatrix} \partial/\partial r \\ \partial/\partial z \\ (1/r)\,\partial/\partial \phi \end{bmatrix}, \quad \text{div} = [1/r + \partial/\partial r \quad \partial/\partial z \quad (1/r)\,\partial/\partial \phi]$$

$$(6.1)$$

It will be noted that the divergence operator contains an extra term $1/r$, which may be thought of as representing the effect of the inherent divergence of the coordinate system. Although in this chapter the symbol ∇ is still used to denote the gradient operator, it is important to realise that in general curvilinear coordinates the symbol ∇^2 denotes div(grad), not $\nabla^t\nabla$. A similar change also appears in the relationship between the two operators associated with elastic stress analysis, which were defined as $\square$ and $\square^t$ in chapter 3.

6.1   Poisson's equation with axial symmetry – the linear ring element
Consider the solution of Poisson's equation

$$D\nabla^2 u = -w \qquad\qquad (6.2)$$

in the axisymmetric region R shown in Fig. 6.1, the source distribution w and the potential u being independent of ϕ. The region R is divided into 'ring' elements, each of triangular cross-section, a typical element being shown in Fig. 6.2. As in section 2.2, the simplest form for an approximating solution $u(r, z)$ is one which is linear within each triangle. Within the triangle shown in Fig. 6.2 this solution may be written in the familiar form

Fig. 6.1. An axisymmetric solution region.

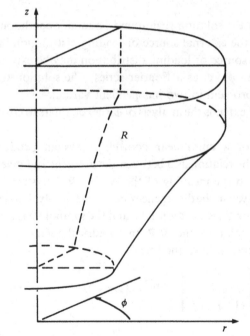

$u = u_i n_i$, where each 'nodal' value u_i represents the value of u at all points on a 'nodal circle' round the z axis. Since the solution is assumed to be independent of the angular coordinate ϕ the operators given in (6.1) become

$$\text{grad} = \nabla = \begin{bmatrix} \partial/\partial r \\ \partial/\partial z \end{bmatrix}, \quad \text{div} = [1/r + \partial/\partial r \quad \partial/\partial z] \tag{6.3}$$

The analysis of section 2.2 may now be carried out as far as equation (2.20a), the only difference being that the integrations are carried out over the *volume* of the solution region R. The contribution of the element of Fig. 6.2 to the nodal equations is

Left-hand side, column j' Right-hand side

$$\text{Add to row } i' \quad \left[2\pi D \int_A \mathbf{b}_i \cdot \mathbf{b}_j \, r \, dA \right] u_{j'} \quad 2\pi \int_A w n_i r \, dA \tag{6.4}$$

$$(i, j = 1, 2, 3)$$

Since the vectors $\mathbf{b}_i$ are constant within A, the left-hand integrals in (6.4) reduce to simple products. The coefficients k_{ij} and w_i in (2.20b) may therefore be written as

$$k_{ij} = DV \mathbf{b}_i \cdot \mathbf{b}_j \tag{6.5a}$$

$$w_i = 2\pi \int_A w n_i r \, dA \quad (i, j = 1, 2, 3) \tag{6.5b}$$

Fig. 6.2. An axisymmetric ring element with linear variation of the dependent variable u.

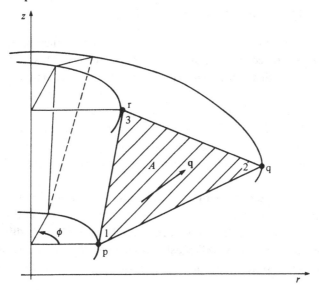

where $V = 2\pi \int_A r \, dA$ is the volume of the ring. The quantity w_i represents the total strength of the equivalent line source uniformly distributed round the circumference of the associated nodal circle. If w is constant over A the total source strength associated with the element is $2\pi w A \bar{r}$, where $\bar{r}$ is the distance of the centroid of the triangular area A from the z axis. The total equivalent line source associated with each nodal circle is approximately $(2\pi w A \bar{r})/3$, provided that $\bar{r}$ is large compared with the dimensions of the element. (This is in contrast to the case of plane potential flow, where the division of the total source strength into three equal equivalent nodal sources is exact if w is constant.)

A numerical solution derived from the above analysis is very similar in character to one obtained from the analysis of section 2.2 for the corresponding plane potential-flow problem. Like the linear plane triangle, the linear ring element gives continuity of value on the inter-element boundaries, and the magnitude of the flow-density approximation $\mathbf{q}$ is constant within each element. There is, however, one important difference. In the two-dimensional solution the flow $\mathbf{q}$ is constant in both magnitude and direction, which implies $\nabla \cdot \mathbf{q} = 0$ within each element – that is, the approximate solution is one in which there is no continuous distribution of sources. In the axisymmetric solution, however, it is only in the r, z plane that the calculated flow $\mathbf{q}$ appears constant within each element. When looked at as a three-dimensional flow pattern the flow diverges, implying a source distribution of density

$$\text{div } \mathbf{q} = [1/r + \partial/\partial r \quad \partial/\partial z]\begin{bmatrix} q_r \\ q_z \end{bmatrix} = q_r/r$$

per unit volume. This result might appear to imply that the approximate solution u is very inaccurate at points on the axis. However, this is not really the case, since a distributed source of density q_r/r per unit *volume* is precisely the same thing as a radial outflow of q_r per unit *area* through the surface of a cylinder whose axis coincides with the axis of symmetry. Of course, if a problem actually involves a finite line-source along the z axis the assumed linear variation of u with r will lead to inaccuracy, since in the vicinity of a line source u varies as $-\log r$ and $\mathbf{q}$ varies as $1/r$.

6.2   Poisson's equation with axial symmetry – higher-order elements

Higher-order axisymmetric ring elements can be constructed according to the procedures set out in chapters 4 and 5. The only difference from the procedure for two-dimensional elements lies in the region of integration, which is now the volume of the ring. In the case of the six-node triangle of section 4.2 the coefficients k_{ij} and w_i are given by a modified version of (4.14),

$$k_{ij} = 2\pi D \iint_A \mathbf{b}_i \cdot \mathbf{b}_j (r_0 + \alpha) \, \mathrm{d}\alpha \, \mathrm{d}\beta, \quad w_i = 2\pi \iint_A w n_i (r_0 + \alpha) \, \mathrm{d}\alpha \, \mathrm{d}\beta$$

(6.6)

where r_0 is the distance from the axis of symmetry to the origin of the local coordinate axes α, β, as shown in Fig. 6.3, and i, j scan over all the nodes associated with the element. The presence of r_0 in (6.6) reflects the fact that the properties of an axisymmetric ring element depend on the distance of the cross-section from the axis of symmetry. This is in contrast to the corresponding planar element, which has the same properties wherever it is placed in the solution region.

The same difference occurs with the elements of more general geometry described in chapter 5. The mapping from the α, β coordinate system to the r, z coordinate system takes place in exactly the same way as for plane elements, the mapping taking the general form

$$\begin{bmatrix} r \\ z \end{bmatrix} = n_k^M (\alpha, \beta) \begin{bmatrix} r_k \\ z_k \end{bmatrix}$$

(6.7)

where k scans over the nodes associated with the mapping and the functions n_k^M are the shape functions associated with those nodes. (As discussed in chapter 5, the functions used for the mapping may be different from those used for the functional approximation.) Equation (6.7) is essentially the same as (5.4), (5.13) or (5.15). The vectors $\mathbf{b}_i(\alpha, \beta) = \nabla n_i$ are still given by (5.8), the coefficients k_{ij} and w_i being given by a modified

Fig. 6.3. The relationship between local and global coordinates for a six-node ring element of fixed geometry.

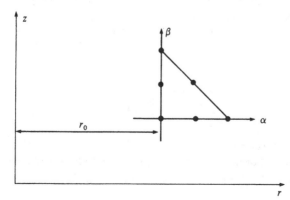

form of (5.9),

$$k_{ij} = 2\pi D \iint_{A'} \mathbf{b}_i \cdot \mathbf{b}_j |\mathbf{J}| \, r(\alpha, \beta) \, d\alpha \, d\beta, \quad w_i = 2\pi \iint_{A'} w n_i |\mathbf{J}| \, r(\alpha, \beta) \, d\alpha \, d\beta$$

(6.8)

The function $r(\alpha, \beta)$ is already available as a by-product of the mapping, being given by the first scalar equation of (6.7),

$$r(\alpha, \beta) = n_k^M(\alpha, \beta) \, r_k$$

(6.9)

The presence of the extra term in the integrand affects the number of Gauss points required for exact evaluation of the integrals in (6.6) and (6.8), but otherwise the numerical integration is carried out in exactly the same way as before.

6.3   Stress analysis with axial symmetry

Chapters 2 and 3 emphasised the similarity between the finite elements used in plane potential problems and the corresponding elements used in plane elastic stress analysis. This similarity also exists in elements with axial symmetry.

An elastic solid under conditions of axial symmetry has four components of strain, which are related to the components of displacement by the equations

$$\varepsilon = \begin{bmatrix} \varepsilon_{rr} \\ \varepsilon_{zz} \\ \varepsilon_{\phi\phi} \\ \gamma_{rz} \end{bmatrix} = \begin{bmatrix} \partial/\partial r & 0 \\ 0 & \partial/\partial z \\ 1/r & 0 \\ \partial/\partial z & \partial/\partial r \end{bmatrix} \begin{bmatrix} u_r \\ u_z \end{bmatrix} \quad \begin{array}{l} \text{(radial)} \\ \text{(axial)} \\ \text{(hoop)} \\ \text{(shear)} \end{array}$$

(6.10a)

Using the notation of equation (3.1), equation (6.10a) is written as

$$\varepsilon = \square \mathbf{u}$$

(6.10b)

Note particularly the presence of the hoop-strain component $\varepsilon_{\phi\phi}$ in (6.10a). This is in contrast to the zero value of the tangential component of flow q_ϕ in an axisymmetric potential problem. The stress vector σ is related to ε by the equation

$$\begin{bmatrix} \sigma_{rr} \\ \sigma_{zz} \\ \sigma_{\phi\phi} \\ \tau_{rz} \end{bmatrix} = c_1 \begin{bmatrix} 1 & c_2 & c_2 & 0 \\ & 1 & c_2 & 0 \\ & & 1 & 0 \\ \text{symmetric} & & & c_3 \end{bmatrix} \begin{bmatrix} \varepsilon_{rr} \\ \varepsilon_{zz} \\ \varepsilon_{\phi\phi} \\ \gamma_{rz} \end{bmatrix}$$

(6.11a)

where $c_1 = E(1-v)/(1+v)(1-2v)$, $c_2 = v/(1-v)$, $c_3 = (1-2v)/2(1-v)$. This equation may be obtained by dropping the last two rows and columns from the full three-dimensional relationship (3.16a). Equation (6.11a) is written in the usual way as

$$\boldsymbol{\sigma} = \mathbf{D}\boldsymbol{\varepsilon} \tag{6.11b}$$

Consideration of the equilibrium of the infinitesimal element shown in Fig. 6.4 gives the equation

$$\begin{bmatrix} 1/r + \partial/\partial r & 0 & -1/r & \partial/\partial z \\ 0 & \partial/\partial z & 0 & 1/r + \partial/\partial r \end{bmatrix} \begin{bmatrix} \sigma_{rr} \\ \sigma_{zz} \\ \sigma_{\phi\phi} \\ \tau_{rz} \end{bmatrix} = -\begin{bmatrix} w_r \\ w_z \end{bmatrix} \tag{6.12}$$

in place of equation (3.3). As with the corresponding equations of potential theory, the change to curvilinear coordinates results in the matrix operator in (6.12) no longer being equal to $\square^t$.

With the symbols re-defined in the above manner the analysis for the axisymmetric linear displacement element follows the pattern set out in section (3.2). Equations (3.14) become

$$\mathbf{K}_{ij} = 2\pi \iint_A \mathbf{B}_i^t \mathbf{D}\mathbf{B}_j r \, dr \, dz, \quad \mathbf{w}_i = 2\pi \iint_A n_i \mathbf{w} r \, dr \, dz \tag{6.13}$$

where

$$\mathbf{B}_i = \square n_i = \begin{bmatrix} \partial n_i/\partial r & 0 \\ 0 & \partial n_i/\partial z \\ n_i/r & 0 \\ \partial n_i/\partial z & \partial n_i/\partial r \end{bmatrix} \tag{6.14}$$

Fig. 6.4. An element of volume showing the surface tractions and body forces which maintain the element in equilibrium.

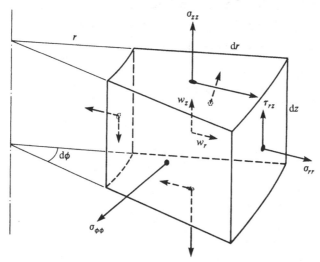

The terms $\partial n_i/\partial r$, $\partial n_i/\partial z$ are exactly the same constant terms as appear in the linear element developed in section 6.1 for use with Poisson's equation. The additional term n_i/r is not constant, which means that the expression for $\mathbf{K}_{ij}$ does not simplify in the same way as the expression for k_{ij} in (6.4).

The treatment of higher-order elements follows a similar pattern to that given in section 6.2, with equation (6.6) being replaced by

$$\mathbf{K}_{ij} = 2\pi \iint_A \mathbf{B}_i^t \mathbf{D}\mathbf{B}_j(r_0+\alpha)\,\mathrm{d}\alpha\,\mathrm{d}\beta, \quad \mathbf{w}_i = 2\pi \iint_A n_i \mathbf{w}(r_0+\alpha)\,\mathrm{d}\alpha\,\mathrm{d}\beta \tag{6.15}$$

and equation (6.8) being replaced by

$$\left.\begin{aligned}\mathbf{K}_{ij} &= 2\pi \iint_{A'} \mathbf{B}_i^t \mathbf{D}\mathbf{B}_j |\mathbf{J}|\, r(\alpha,\beta)\,\mathrm{d}\alpha\,\mathrm{d}\beta \\ \mathbf{w}_i &= 2\pi \iint_{A'} n_i \mathbf{w}|\mathbf{J}|\, r(\alpha,\beta)\,\mathrm{d}\alpha\,\mathrm{d}\beta\end{aligned}\right\} \tag{6.16}$$

6.4   The calculation of equivalent nodal sources and loads

As mentioned in section 6.1, the quantity w_i in (6.5b) represents the total strength of a line source uniformly distributed round the circumference of a nodal circle. This fact must be borne in mind when calculating the nodal sources or nodal loads associated with a specified axisymmetric surface distribution of flow or load.

As an example, consider a uniform distribution of surface loading w per unit area on the annulus formed by one face of a quadratic ring element, as shown in Fig. 6.5. The shape functions associated with the quadratic displacement variation are still the functions

$$n_1 = (\alpha-1)\alpha/2, \quad n_2 = 1-\alpha^2, \quad n_3 = (1+\alpha)\alpha/2$$

illustrated in Fig. 1.1. However, the virtual work equations (4.12) must be

Fig. 6.5. Uniform loading on an annular ring with quadratic variation of displacement.

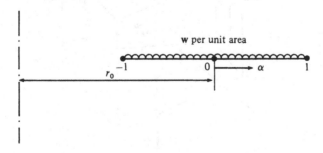

replaced by the equations

$$\mathbf{w}_i = 2\pi \int_{-1}^{1} n_i \mathbf{w}(r_0 + \alpha) \, d\alpha \tag{6.17}$$

which gives

$$\mathbf{w}_1 = (\mathbf{W}/6)(r_0 - 1)/r_0$$
$$\mathbf{w}_2 = 2\mathbf{W}/3$$
$$\mathbf{w}_3 = (\mathbf{W}/6)(r_0 + 1)/r_0$$

where $\mathbf{W} = 4\pi \mathbf{w} r_0$ is the total surface load. These expressions should be compared with those obtained in section 4.7 for the quadratic plane element.

6.5   Harmonic solutions of field problems

Cylindrical polar coordinates are the natural coordinates for the analysis of any region with *geometrical* axial symmetry, even under conditions where the solution itself is not symmetric. Such cases can be analysed most easily by using ring elements in which the approximating functions vary harmonically with ϕ, the solution being expressed as a Fourier series.

As an example, consider the solution of Poisson's equation

$$D\nabla^2 u = -w \tag{6.18}$$

in the axisymmetric region shown in Fig. 6.1, w being a function of r, z and ϕ. To simplify the example, assume that w is symmetric about the plane $\phi = 0$,† so that it can be written in the form

$$w(r, z, \phi) = \sum_k w^{(k)}(r, z) \cos k\phi \quad (k = 0, 1, 2, \ldots) \tag{6.19}$$

The functions $w^{(k)}(r, z)$ are determined from the Fourier relationships

$$\left. \begin{aligned} w^{(0)}(r, z) &= (1/\pi) \int_0^\pi w(r, z, \phi) \, d\phi \\ w^{(k)}(r, z) &= (2/\pi) \int_0^\pi w(r, z, \phi) \cos k\phi \, d\phi \end{aligned} \right\} \tag{6.20}$$

Since (6.18) is a linear equation its solution is the sum of the solutions associated with the individual terms in (6.19), the solution associated with the k'th harmonic $w^{(k)}(r, z) \cos k\phi$ being of the form $u^{(k)}(r, z) \cos k\phi$. Within each ring element the polynomial approximation to this component of the solution is written in terms of a set of nodal values $u_i^{(k)}$ as $u_i^{(k)} n_i(r, z) \cos k\phi$.

---

† This assumption merely removes the terms in $\sin k\phi$, which are treated in a similar manner to those in $\cos k\phi$.

The analysis for each harmonic solution now follows a similar pattern to that set out in sections 6.1 and 6.2, with ∇ taking its full three-dimensional form (equation (6.1)) and n_i and w being replaced by $n_i(r,z) \cos k\phi$ and $w^{(k)}(r,z) \cos k\phi$ respectively. The coefficients $k_{ij}^{(k)}$ are given by

$$k_{ij}^{(k)} = D \int_0^{2\pi} \int_A \mathbf{b}_i^{(k)} \cdot \mathbf{b}_j^{(k)} \, r \, \mathrm{d}A \, \mathrm{d}\phi \tag{6.21}$$

where

$$\mathbf{b}_i^{(k)} = \nabla(n_i \cos k\phi) = \begin{bmatrix} \dfrac{\partial n_i}{\partial r} \cos k\phi \\[2mm] \dfrac{\partial n_i}{\partial z} \cos k\phi \\[2mm] -\dfrac{kn_i}{r} \sin k\phi \end{bmatrix}$$

Each term in the products $\mathbf{b}_i^{(k)} \cdot \mathbf{b}_j^{(k)}$ in (6.21) has a multiplier $\cos^2 k\phi$ or $\sin^2 k\phi$, so that the integration with respect to ϕ is trivial. The result may be written in the form,

$$k_{ij}^{(k)} = \pi D \int_A \bar{\mathbf{b}}_i^{(k)} \cdot \bar{\mathbf{b}}_j^{(k)} \, r \, \mathrm{d}A \tag{6.22}$$

where

$$\bar{\mathbf{b}}_i^{(k)} = \begin{bmatrix} \dfrac{\partial n_i}{\partial r} \\[2mm] \dfrac{\partial n_i}{\partial z} \\[2mm] -\dfrac{kn_i}{r} \end{bmatrix}$$

Note that $\mathbf{b}_i^{(k)}$ is a true vector function in r, z, ϕ space, while $\bar{\mathbf{b}}_i^{(k)}$ is simply a triad of coefficients written as a vector to make (6.22) resemble (6.21). If $k = 0$ and the shape functions n_i are linear, equation (6.22) reduces to (6.5a). If the element has node p on the z axis then the component kn_p/r in $\bar{\mathbf{b}}_p^{(k)}$ produces an integrand of order $1/r$ in (6.22). However, if $k \neq 0$ and u is single-valued on the z axis then the nodal value $u_p^{(k)}$ is necessarily zero, so that the associated (infinite) coefficient $k_{pp}^{(k)}$ does not appear in the nodal equations.

In the same way the coefficients $w_i^{(k)}$ are given by

$$w_i^{(k)} = \int_0^{2\pi} \int_A w^{(k)} n_i \cos^2(k\phi) \, r \, \mathrm{d}A \, \mathrm{d}\phi$$

which integrates to give

$$w_i^{(k)} = \pi \int_A w^{(k)} n_i \, r \, \mathrm{d}A \tag{6.23}$$

If necessary the integrals in (6.22) and (6.23) may be evaluated by transforming from r, z to α, β variables in the manner described in section 6.2.

The treatment of the analogous problem in stress analysis follows a similar pattern. Equation (6.10a) is replaced by the complete three-dimensional relationship

$$\varepsilon = \begin{bmatrix} \varepsilon_{rr} \\ \varepsilon_{zz} \\ \varepsilon_{\phi\phi} \\ \gamma_{rz} \\ \gamma_{r\phi} \\ \gamma_{z\phi} \end{bmatrix} = \begin{bmatrix} \partial/\partial r & 0 & 0 \\ 0 & \partial/\partial z & 0 \\ 1/r & 0 & (1/r)\partial/\partial\phi \\ \partial/\partial z & \partial/\partial r & 0 \\ (1/r)\partial/\partial\phi & 0 & \partial/\partial r - 1/r \\ 0 & (1/r)\partial/\partial\phi & \partial/\partial z \end{bmatrix} \begin{bmatrix} u_r \\ u_z \\ u_\phi \end{bmatrix} \tag{6.24a}$$

which may be written in the usual way as

$$\varepsilon = \square \mathbf{u} \tag{6.24b}$$

If the loading $\mathbf{w}$ is symmetric about the plane $\phi = 0$ then the equation which corresponds to (6.19) is

$$\mathbf{w}(r, z, \phi) = \sum_k \begin{bmatrix} w_r^{(k)}(r, z) \cos k\phi \\ w_z^{(k)}(r, z) \cos k\phi \\ w_\phi^{(k)}(r, z) \sin k\phi \end{bmatrix}$$

(Note the change from $\cos k\phi$ to $\sin k\phi$ in the tangential component of load.) The displacement approximation $\mathbf{u}$ takes a similar form, the k'th harmonic being given by

$$\mathbf{u}^{(k)}(r, z, \phi) = \begin{bmatrix} (u_r^{(k)})_i \, n_i(r, z) \cos k\phi \\ (u_z^{(k)})_i \, n_i(r, z) \cos k\phi \\ (u_\phi^{(k)})_i \, n_i(r, z) \sin k\phi \end{bmatrix}$$

Equation (6.21) becomes

$$\mathbf{K}_{ij}^{(k)} = \int_0^{2\pi} \iint_A (\mathbf{B}_i^{(k)})^t \mathbf{D} \mathbf{B}_j^{(k)} \, r \, \mathrm{d}r \, \mathrm{d}z \, \mathrm{d}\phi \tag{6.25}$$

where

$$\mathbf{B}_i^{(k)} = \Box \begin{bmatrix} n_i \cos k\phi & 0 & 0 \\ 0 & n_i \cos k\phi & 0 \\ 0 & 0 & n_i \sin k\phi \end{bmatrix}$$

$$= \begin{bmatrix} \dfrac{\partial n_i}{\partial r} \cos k\phi & 0 & 0 \\[2mm] 0 & \dfrac{\partial n_i}{\partial z} \cos k\phi & 0 \\[2mm] \dfrac{n_i}{r} \cos k\phi & 0 & \dfrac{kn_i}{r} \cos k\phi \\[2mm] \dfrac{\partial n_i}{\partial z} \cos k\phi & \dfrac{\partial n_i}{\partial r} \cos k\phi & 0 \\[2mm] \dfrac{-kn_i}{r} \sin k\phi & 0 & \left(\dfrac{\partial n_i}{\partial r} - \dfrac{n_i}{r}\right) \sin k\phi \\[2mm] 0 & \dfrac{-kn_i}{r} \sin k\phi & \dfrac{\partial n_i}{\partial z} \sin k\phi \end{bmatrix} \quad (6.26)$$

and $\mathbf{D}$ is given by (3.16a). As in harmonic solutions of Poisson's equation, each term in the product $(\mathbf{B}_i^{(k)})^t \, \mathbf{D} \mathbf{B}_j^{(k)}$ has a multiplier $\cos^2 k\phi$ or $\sin^2 k\phi$, so that integration with respect to ϕ is trivial. The result may be written in the form

$$\mathbf{K}_{ij}^{(k)} = \pi \iint_A (\bar{\mathbf{B}}_i^{(k)})^t \mathbf{D} \bar{\mathbf{B}}_j^{(k)} \, r \, dr \, dz \qquad (6.27)$$

the matrices $\bar{\mathbf{B}}_i^{(k)}$ being given by equation (6.26) with the multipliers $\cos k\phi$ and $\sin k\phi$ omitted. The treatment of the loading $\mathbf{w}$ follows a similar pattern.

It is also possible to use an approximating function which is a Fourier series in a linear coordinate variable. For example, if a box-girder bridge of span L has constant cross-section it is appropriate to write the displacement $\mathbf{u}$ in the form

$$\mathbf{u}(x, y, z) = \sum_k [\mathbf{u}_s^{(k)}(x, y) \sin 2k\pi z/L + \mathbf{u}_c^{(k)}(x, y) \cos 2k\pi z/L] \quad (6.28)$$

where z is the longitudinal axis of the bridge. The loading $\mathbf{w}$ may be expressed in a similar manner. The cross-section can then be divided into a set of prismatic elements and the analysis carried out as a series of two-dimensional problems in the x, y plane.

Sometimes there is a choice of harmonic variable, as in the cylinder shown in Fig. 6.6. In Fig. 6.6a the harmonic variable is ϕ and each harmonic solution involves four ring elements derived from the eight-node square element of section 4.4. In Fig. 6.6*b* the harmonic variable is *z* and each harmonic solution involves four eight-node isoparametric prismatic elements. In the latter case the boundaries of the distorted square elements only represent an approximation to the true boundary of the cylinder.

Problems for chapter 6

6.1 Find the stiffness coefficients for an axisymmetric annular disc of constant thickness d, internal radius r_1 and external radius r_2, subjected to uniform radial forces on its inner and outer edges. Assume that the radial displacement u is a linear function of the radius r and that the disc is sufficiently thin for the material to be in a state of plane stress.

6.2 Generalise the analysis of section 6.4 to find the equivalent nodal loads for the case when the loaded face of the quadratic ring element is a cone of semi-angle ψ.

6.3 The ring element of square cross-section shown in Fig. 6.7 is part of a body rotating about the z axis with angular velocity ω, so that the

Fig. 6.6. Two finite-element arrangements for the analysis of a circular cylinder.
(*a*) using harmonics in ϕ,
(*b*) using harmonics in z.

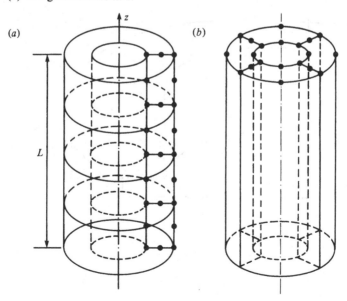

material of the element is subject to a radial body-force equal to $\rho\omega^2 r$ per unit volume. What are the equivalent nodal loads associated with the element?

6.4 Fig. 6.8 shows a prismatic element of constant cross-section which is to be used in a finite-element solution of Poisson's equation $D\nabla^2 u = -w$ within a prismatic bar. The dependent variable $u(x, y, z)$ is written in the

form $\sum\limits_{k} u^{(k)}(x, y) \sin k\pi z/L$, the source distribution w being expressed in a similar manner. The two-dimensional distribution $u^{(k)}(x, y)$ is replaced by an approximation $u_i^{(k)} n_i(x, y)$ which is linear within the triangular cross-section of the element. Obtain expressions for the nodal coefficients $k_{ij}^{(k)}$ and the equivalent nodal sources $w_i^{(k)}$ for the element. [A result derived as part of the solution to problem 3.7 is useful here.]

6.5 Fig. 6.8 may also be thought of as representing an element in a prismatic beam of arbitrary cross-section under some distribution of transverse load w. Find suitable harmonic expansions for the components of $\mathbf{u}$ and $\mathbf{w}$ if the boundary conditions at the two ends of the beam are those of 'simple support' (i.e. $u_x = u_y = 0$, $\sigma_{zz} = 0$). Repeat the analysis of problem 6.4 to obtain the element stiffness matrices $\mathbf{K}_{ij}^{(k)}$ and the equivalent nodal loads $\mathbf{w}_i^{(k)}$ for the k'th harmonic, assuming $\mathbf{u}^{(k)}(x, y)$ to be replaced by an approximation which is linear within the triangle.

Fig. 6.7.

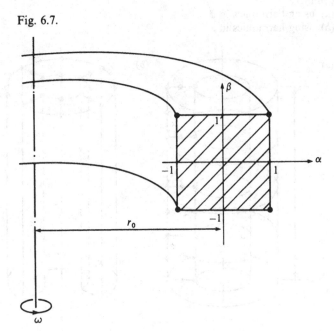

Solutions to problems

6.1 It is convenient to use the notation of Fig. 6.9, writing $r = r_0 + \alpha$, $-L/2 \leqslant \alpha \leqslant L/2$, where $r_0 = (r_1 + r_2)/2$, $L = r_2 - r_1$. The radial displacement u is assumed to be of the form $u = u_i n_i$, where $n_i = (L/2 + \alpha_i \alpha)/L$, $\alpha_1 = -1$, $\alpha_2 = 1$. The strain vector ε is given by

$$\varepsilon = \begin{bmatrix} \varepsilon_{rr} \\ \varepsilon_{\phi\phi} \end{bmatrix} = \begin{bmatrix} \mathrm{d}/\mathrm{d}r \\ 1/r \end{bmatrix} u = \begin{bmatrix} \mathrm{d}n_i/\mathrm{d}r \\ n_i/r \end{bmatrix} u_i = \begin{bmatrix} \alpha_i/L \\ (L/2 + \alpha_i \alpha)/L(r_0 + \alpha) \end{bmatrix} u_i = \mathbf{b}_i u_i$$

Since the disk is thin the matrix $\mathbf{D}$ takes its 'plane stress' form

$$\mathbf{D} = \frac{E}{1 - v^2} \begin{bmatrix} 1 & v \\ v & 1 \end{bmatrix}$$

Fig. 6.8.

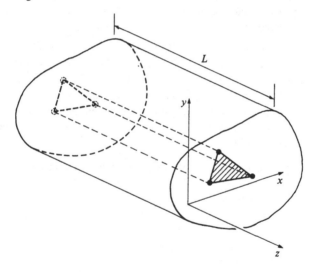

Fig. 6.9.

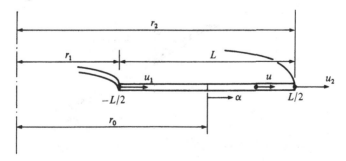

The matrices $\mathbf{K}_{ij}$ are single coefficients, given by

$$K_{ij} = 2\pi d \int_{-L/2}^{L/2} \mathbf{b}_i^t \mathbf{D} \mathbf{b}_j (r_0 + \alpha)\, d\alpha$$

The evaluation of this expression is somewhat tedious, the result being

$$K_{ij} = \frac{2\pi d E}{1 - v^2} [\alpha_i \alpha_j r_0 / L + (1 + v)(\alpha_i + \alpha_j)/2$$
$$+ \{1/4 - r_0(\alpha_i + \alpha_j)/2L + \alpha_i \alpha_j r_0^2 / L^2\} \log (r_2/r_1)]$$

If reduced (one-point) integration is used, the corresponding result is

$$K_{ij} = \frac{2\pi d E}{1 - v^2} [\alpha_i \alpha_j r_0 / L + v(\alpha_i + \alpha_j)/2 + L/4r_0]$$

6.2 Using the notation of Fig. 6.10, the load on an annular ring of width $d\alpha$ is $w2\pi r\, d\alpha$, where $r = r_0 + \alpha \sin \psi$. Hence

$$\mathbf{w}_i = 2\pi \int_{-1}^{1} n_i wr\, d\alpha = 2\pi \int_{-1}^{1} n_i w(r_0 + \alpha \sin \psi)\, d\alpha$$

If w is constant and the functions n_i are the quadratic shape functions defined in section 6.4, straightforward integration gives

$$\mathbf{w}_1 = \mathbf{W}(r_0 - \sin \psi)/6r_0, \quad \mathbf{w}_2 = 2\mathbf{W}/3, \quad \mathbf{w}_3 = \mathbf{W}(r_0 + \sin \psi)/6r_0$$

where $\mathbf{W}$ is the total surface load, equal to $4\pi w r_0$. (These expressions are independent of the direction of w). Putting $\psi = \pi/2$ gives the expressions in section 6.4. Putting $\psi = 0$ gives the expressions in section 4.7. These expressions also hold for a straight element boundary in which the range of α is $\pm L/2$, provided that r_0 is interpreted as a non-dimensional radius, equal to $(r_1 + r_2)/2L$.

6.3 The rotation produces an outwards radial inertia loading $\rho \omega^2 (r_0 + \alpha)$ per

Fig. 6.10.

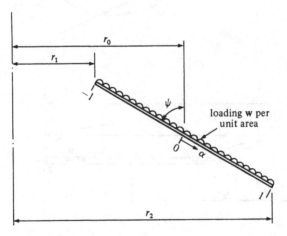

loading w per
unit area

unit volume, where r_0 is the distance from the axis of symmetry to the origin of local coordinates. The radial components of the equivalent nodal loads are therefore

$$(w_r)_i = 2\pi\rho\omega^2 \int_{-1}^{1}\int_{-1}^{1} n_i(r_0+\alpha)^2 \, d\alpha \, d\beta$$

where the shape functions n_i are those given in equation (4.4). Straight-forward integration gives

$$(w_r)_i = Mr_0\omega^2(3r_0^2+2\alpha_i r_0+1)/12$$

where $M = 8\pi\rho r_0$ is the mass of the element. The axial components of the equivalent nodal loads are, of course, zero.

6.4 If $u(x, y, z) = \sum_k u^{(k)}(x, y) \sin k\pi z/L$, $(k = 1, 2, \ldots)$ and $u^{(k)}(x, y) = u_i^{(k)} n_i(x, y)$ then

$$\mathbf{b}_i^{(k)} = \nabla(n_i \sin k\pi z/L) = \begin{bmatrix} \dfrac{\partial n_i}{\partial x} \sin k\pi z/L \\[2mm] \dfrac{\partial n_i}{\partial y} \sin k\pi z/L \\[2mm] \dfrac{k\pi n_i}{L} \cos k\pi z/L \end{bmatrix}$$

and

$$k_{ij}^{(k)} = D \int_0^L \int_A \mathbf{b}_i^{(k)} \cdot \mathbf{b}_j^{(k)} \, dA \, dz$$

Integration with respect to z gives

$$k_{ij}^{(k)} = LDA\mathbf{b}_i \cdot \mathbf{b}_j/2 + (k^2\pi^2 D/2L) \int_A n_i n_j \, dA$$

where $\mathbf{b}_i$ is the vector ∇n_i defined in section 2.1. Using the results derived in the solution to problem 3.7, this expression may be written as

$$k_{ij}^{(k)} = k_{ij}^{(0)}/2 + k^2\pi DA/12L \quad (i = j)$$
$$= k_{ij}^{(0)}/2 + k^2\pi DA/24L \quad (i \neq j)$$

where $k_{ij}^{(0)}$ is the corresponding coefficient for a normal plane element of thickness L. The equivalent nodal sources are

$$w_i^{(k)} = \int_0^L \int_A w^{(k)} n_i \sin^2(k\pi z/L) \, dA \, dz = (L/2) \int_A w^{(k)} n_i \, dA$$

If $w^{(k)}$ is constant this expression reduces to $w^{(k)} LA/6$ for all values of i.

6.5 If $u_x = u_y = 0$ at the ends of the beam it is appropriate to introduce the series expansion

$$u_x(x, y, z) = \sum_k u_x^{(k)}(x, y) \sin k\pi z/L \quad (k = 1, 2, \ldots)$$

with a similar expansion for u_y. These expansions imply $\varepsilon_{zz} = \partial u_z/\partial x = 0$

and $\varepsilon_{yy} = \partial u_y / \partial y = 0$ when $z = 0, L$. It follows that the stress condition $\sigma_{zz} = 0$ at $z = 0, L$ implies $\varepsilon_{zz} = 0$ at $z = 0, L$. Thus an appropriate expansion for u_z is

$$u_z(x, y, z) = \sum_k u_z^{(k)}(x, y) \cos k\pi z / L$$

The components of the transverse loading **w** may similarly be expressed as

$$w_x(x, y, z) = \sum_k w_x^{(k)}(x, y) \sin k\pi z / L$$

$$w_y(x, y, z) = \sum_k w_y^{(k)}(x, y) \sin k\pi z / L$$

$$w_z(x, y, z) = 0$$

[Note that a general distribution of transverse load can always be represented by the half-range sine series given above. There is no need to introduce cosine terms in order to achieve more 'generality', since there are no boundary conditions on **w** or its derivatives.]

The harmonic component of displacement $\mathbf{u}^{(k)}(x, y, z)$ may be written as

$$\mathbf{u}^{(k)}(x, y, z) = \begin{bmatrix} (u_x^{(k)})_i n_i(x, y) \sin k\pi z / L \\ (u_y^{(k)})_i n_i(x, y) \sin k\pi z / L \\ (u_z^{(k)})_k n_i(x, y) \cos k\pi z / L \end{bmatrix}$$

Equation (6.25) becomes

$$\mathbf{K}_{ij}^{(k)} = \int_0^L \int_A (\mathbf{B}_i^{(k)})^t \, \mathbf{D} \mathbf{B}_j^{(k)} \, \mathrm{d}A \, \mathrm{d}z \qquad (6.29)$$

where

$$\mathbf{B}_i^{(k)} = \square \begin{bmatrix} n_i \sin k\pi z / L & 0 & 0 \\ 0 & n_i \sin k\pi z / L & 0 \\ 0 & 0 & n_i \cos k\pi z / L \end{bmatrix}$$

$\square$ being the stress operator for three-dimensional rectangular Cartesian coordinates given in (3.15a). Thus

$$\mathbf{B}_i^{(k)} = \begin{bmatrix} \dfrac{\partial n_i}{\partial x} \sin k\pi z / L & 0 & 0 \\[2mm] 0 & \dfrac{\partial n_i}{\partial y} \sin k\pi z / L & 0 \\[2mm] 0 & 0 & \dfrac{-k\pi n_i}{L} \sin k\pi z / L \\[2mm] \dfrac{\partial n_i}{\partial y} \sin k\pi z / L & \dfrac{\partial n_i}{\partial x} \sin k\pi z / L & 0 \\[2mm] 0 & \dfrac{k\pi n_i}{L} \cos k\pi z / L & \dfrac{\partial n_i}{\partial y} \cos k\pi z / L \\[2mm] \dfrac{k\pi n_i}{L} \cos k\pi z / L & 0 & \dfrac{\partial n_i}{\partial x} \cos k\pi z / L \end{bmatrix} \qquad (6.30)$$

The constitutive matrix $\mathbf{D}$ is given by (3.16a). As in the corresponding analysis in section 6.5, each term in the product $(\mathbf{B}_i^{(k)})^t \mathbf{D} \mathbf{B}_j^{(k)}$ has a multiplier $\sin^2 k\pi z/L$ or $\cos^2 k\pi z/L$, so that integration of (6.29) with respect to z is trivial. The result may be written in the form

$$\mathbf{K}_{ij}^{(k)} = (L/2) \int_A (\bar{\mathbf{B}}_i^{(k)})^t \mathbf{D} \bar{\mathbf{B}}_j^{(k)} \, \mathrm{d}A$$

the matrices $\bar{\mathbf{B}}_i^{(k)}$ being given by expression (6.30) with the multipliers $\sin k\pi z/L$ and $\cos k\pi z/L$ omitted. All the terms in the matrix product are constants, multiples of n_i or multiples of $n_i n_j$. Since $\int_A n_i \, \mathrm{d}A$ and $\int_A n_i n_j \, \mathrm{d}A$ are known quantities (see chapter 2 and the solution to problem 3.7), the elements of the matrices $\mathbf{K}_{ij}$ may be evaluated analytically. If numerical integration is used, four-point integration is exact, although reduced integration may give better results.

The treatment of the loading $\mathbf{w}$ follows a similar pattern, the equivalent nodal loads being given by

$$\mathbf{w}_i^{(k)} = \int_0^L \int_A \mathbf{w}^{(k)} n_i \sin^2 (k\pi z/L) \, \mathrm{d}A \, \mathrm{d}z = (L/2) \int_A \mathbf{w}^{(k)} n_i \, \mathrm{d}A$$

7

The elastic analysis of beams, plates and shells

One of the most important areas of application of the finite-element method during the last decade has been the stress-analysis of structures such as oil-rigs, aircraft, ships and vehicle bodies, which are fabricated from plates and shells. The geometrical complexity of these structures is such that a complete stress analysis may need several thousand elements to achieve adequate accuracy.

In plate and shell elements the nodes lie in a 'middle surface', the 'thickness' normal to the middle surface being small compared with the other dimensions, so that these elements are, in a sense, two-dimensional. A *plate* element is one in which the middle surface is a *plane*. In such elements the stretching and bending modes of deformation are uncoupled and may be treated separately, the stretching being a plane-stress problem which can be treated by the methods described earlier in this book. A *shell* element is one in which the middle surface is *curved*. The curvature has the effect of coupling the two deformation modes – a coupling which gives shells much greater transverse stiffness than plates.

There are two sets of assumptions commonly used in the development of plate and shell elements. The first set allows for shear deformations and leads to 'thick plate' and 'thick shell' elements. The second excludes shear deformations and generates 'thin plate' and 'thin shell' elements. It is the latter set of assumptions which forms the basis of the theory normally used by engineers. Although the equations based on this theory are simpler than either the general continuum equations or the thick-plate and thick-shell equations, their use in the development of finite elements leads to problems of continuity on inter-element boundaries. As mentioned in section 1.3, conforming thin-plate elements require continuity of boundary slope as well as of displacement, and this is difficult to achieve. Indeed, it was the

140

difficulty of constructing conforming elements for thin plates which led to the development of the patch test described in section 5.6 and the various non-conforming elements which satisfy that test.

As an introduction to the development of plate and shell elements, the chapter begins with an account of some simple beam elements. Apart from their value in the analysis of skeletal structures, these elements may be used to solve plate problems in which deflections and stresses are independent of one of the coordinates in the plane of the plate – such problems are essentially ones of plane strain. Beam elements may also be modified to give elements suitable for analysing axisymmetric plates, or plates in which variation with respect to one space coordinate has been removed by a series expansion of the type described in section 6.5.

The chapter continues with an account of some simple plate elements. Apart from their use in modelling flat plates these elements may be used to analyse 'folded plate' structures, where not all the elements lie in the same plane. In such cases the nodal stiffness matrices of the elements must be transformed into a common global coordinate system before the assembly process – a procedure which is also required when beam elements are used to solve skeletal structures such as frames and trusses. The final section considers some simple shell elements.

A great variety of elements have been developed for plate and shell analysis and the latter part of the chapter should be regarded simply as a preparation for study of the standard texts.

7.1 A simple element for the analysis of deep beams in plane bending

In this section a simple element of constant rectangular cross-section is developed for the analysis of deep beams. (A 'deep' beam, like a 'thick' plate, is one in which shear deformations are significant.) This element is a modification of the rectangular four-node element shown in Fig. 7.1, which was introduced in square form in section 4.3.

When the element shown in the figure is used as a plane stress element it has eight nodal degrees of freedom, the displacement approximation $\mathbf{u}(x, z)$ varying linearly on all lines parallel to the x and z axes. The transformation into a beam element involves the following additional approximations,

(*a*) Points on the neutral axis of the beam (i.e. the x axis) are assumed to displace only in the z direction. This implies that there is no overall axial displacement of any cross-section.

(*b*) The external loading is assumed to consist of a distributed transverse load $w(x)$ per unit length acting at the neutral axis.

(c) Transverse displacements relative to the neutral axis are assumed to be small compared with the displacement of the neutral axis.

(d) The transverse stress σ_{zz} is assumed to be zero.

These assumptions allow the displacement approximation $\mathbf{u}(x,z)$ to be expressed in terms of two linearly independent functions of x only – the transverse displacement u of the neutral axis and the rotation θ of lines originally perpendicular to the neutral axis. These quantities, which are shown in Fig. 7.2,† are assumed to be linear in x and are expressed in terms of their values at the two ends of the element in the usual way. Thus the two components of the displacement approximation are given by

$$u_x(x,z) = -z\,\theta(x) = -z\,\theta_i n_i(x), \quad u_z(x,z) = u(x) = u_i n_i(x)$$
$$(i = 1, 2)$$

where the one-dimensional shape functions n_i have the linear form $n_1 = (L/2 - x)/L$, $n_2 = (L/2 + x)/L$. These equations may be written as

$$\begin{bmatrix} u_x \\ u_z \end{bmatrix} = \begin{bmatrix} 0 & -zn_i \\ n_i & 0 \end{bmatrix}\begin{bmatrix} u_i \\ \theta_i \end{bmatrix} \tag{7.1}$$

In this displacement approximation all straight lines initially parallel to the x and z axes remain straight, as in the original four-node element. Thus 'plane sections remain plane', but not necessarily perpendicular to the neutral axis. The nodal loads which correspond to the nodal displacement components u_i and θ_i consist of the forces w_i and moments m_i shown in Fig. 7.2. Note that the modified element has four nodal degrees of freedom in place of the original eight.

The relevant strains are

$$\varepsilon = \begin{bmatrix} \varepsilon_{xx} \\ \gamma_{xz} \end{bmatrix} = \begin{bmatrix} du_x/dx \\ du_z/dx + du_x/dz \end{bmatrix} = \begin{bmatrix} 0 & -z\,dn_i/dx \\ dn_i/dx & -n_i \end{bmatrix}\begin{bmatrix} u_i \\ \theta_i \end{bmatrix}$$

† Note that θ is taken as positive when *anticlockwise* about the y axis. This change from the usual clockwise convention makes it easier to compare the analysis with the theory presented in the next section, in which θ becomes equal to du/dx.

Fig. 7.1. A four-node rectangular plane stress element used to represent a section of a deep beam.

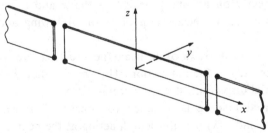

which may be written in the usual way as $\varepsilon = \mathbf{B}_i \mathbf{u}_i$, where

$$\mathbf{B}_1 = \begin{bmatrix} 0 & z/L \\ -1/L & -(L/2-x)/L \end{bmatrix}, \quad \mathbf{B}_2 = \begin{bmatrix} 0 & -z/L \\ 1/L & -(L/2+x)/L \end{bmatrix}$$

$$\mathbf{u}_i = \begin{bmatrix} u_i \\ \theta_i \end{bmatrix} \tag{7.2}$$

The stress/strain relationship is written in the form

$$\boldsymbol{\sigma} = \begin{bmatrix} \sigma_{xx} \\ \tau_{xz} \end{bmatrix} = \begin{bmatrix} E & 0 \\ 0 & G \end{bmatrix} \begin{bmatrix} \varepsilon_{xx} \\ \gamma_{xz} \end{bmatrix} = \mathbf{D}\boldsymbol{\varepsilon} \tag{7.3}$$

where $G = E/2(1+v)$.† The $\mathbf{K}_{ij}$ matrices may now be evaluated from the relationship

$$\mathbf{K}_{ij} = b \int_{-d/2}^{d/2} \int_{-L/2}^{L/2} \mathbf{B}_i^t \mathbf{D} \mathbf{B}_j \, dx \, dz \quad (i,j=1,2) \tag{7.4}$$

The equivalent nodal loads due to the transverse distributed load $w(x)$ are given by

$$w_i = \int_{-L/2}^{L/2} n_i w \, dx, \quad m_i = 0 \tag{7.5}$$

Since the functions n_i are linear, the values of w_i are simply the two concentrated forces statically equivalent to $w(x)$.

† It has been found that the performance of this element is improved if the shear modulus G is multiplied by the factor $\frac{5}{6}$ to allow for the fact that the distribution of shear stress over the cross-section is parabolic rather than constant.

Fig. 7.2. The element of Fig. 7.1 transformed into a two-node deep-beam element.

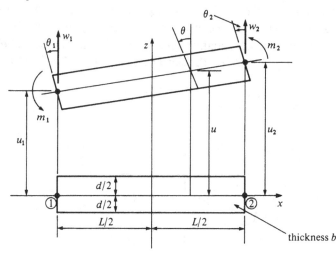

It is convenient to divide the integrand in (7.4) into two parts: 'bending' terms (those multiplied by E) and 'shearing' terms (those multiplied by G). Both sets of integrations are straightforward. The result can also be divided into two parts,

$$\mathbf{K}_{ij} = (\mathbf{K}_{ij})_b + (\mathbf{K}_{ij})_s$$

and each set of 2×2 matrices arranged to form a 4×4 matrix,

$$(\mathbf{K}_{ij})_b = (Ebd^3/12L) \begin{bmatrix} \begin{bmatrix} 0 & 0 \\ 0 & 1 \end{bmatrix} & \begin{bmatrix} 0 & 0 \\ 0 & -1 \end{bmatrix} \\ \begin{bmatrix} 0 & 0 \\ 0 & -1 \end{bmatrix} & \begin{bmatrix} 0 & 0 \\ 0 & 1 \end{bmatrix} \end{bmatrix} \tag{7.6}$$

$$(\mathbf{K}_{ij})_s = (Gbd/L) \begin{bmatrix} \begin{bmatrix} 1 & L/2 \\ L/2 & L^2/3 \end{bmatrix} & \begin{bmatrix} -1 & L/2 \\ -L/2 & L^2/6 \end{bmatrix} \\ \begin{bmatrix} -1 & -L/2 \\ L/2 & L^2/6 \end{bmatrix} & \begin{bmatrix} 1 & -L/2 \\ -L/2 & L^2/3 \end{bmatrix} \end{bmatrix} \tag{7.7a}$$

Note that $\mathbf{K}_{ij}$ has rank 2. This corresponds to the fact that the element in Fig. 7.2 has two rigid-body displacement modes.

The new element, like the four-node continuum element from which it is derived, gives continuity of displacement across inter-element boundaries, so that analyses which use it will converge to the exact solution as the element length is reduced. (The 'exact solution' here means the solution of the 'thick beam' differential equation, which incorporates the approximations made at the beginning of this section.)

Unfortunately, however, the element as so far developed has exactly the same defect as the original four-node square. As discussed in section 4.3, the linear displacement approximation (7.1) does not allow the element to 'bend' properly, the deformation under a pair of end-moments being as shown in Fig. 7.3. As a result the element is too 'stiff'. This is particularly

Fig. 7.3. The deformation of the deep-beam element under equal and opposite end-moments.

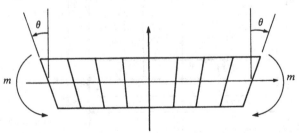

the case when d is small, since if $L \gg d$ the coefficients of $(\mathbf{K}_{ij})_\mathrm{s}$ are very much larger than the coefficients of $(\mathbf{K}_{ij})_\mathrm{b}$.

The performance of the element can be improved greatly by the addition of a node at the origin. This allows the transverse displacement of the neutral axis to vary quadratically with x. Since the element is only connected to other elements at its ends the additional node is an 'internal' node, so that the associated nodal variables may be eliminated from the nodal equations *before* the element assembly process. This results in a two-node element in which the top and bottom edges can become curved. No continuity problems are caused by this, since no displacement compatibility conditions are imposed on the top and bottom edges of a beam.

Essentially the same result can be obtained more economically by the use of reduced integration. The excessive stiffness of the element in the deformation mode of Fig. 7.3 is due to the contribution of the spurious shear strains to the total strain energy. If this contribution is removed then the deformation pattern shown in the figure, though not a correct representation of pure bending, will give the correct relationship between end-moments and end-rotations.

Analytical integration of (7.4) with respect to z gives an integral with x as independent variable in which the 'bending' terms of the integrand are constant and the 'shearing' terms are quadratic in x. It follows that for the bending terms one-point Gauss integration with respect to x gives the exact result (7.6), while for the shearing terms two-point Gauss integration is needed to give the exact result (7.7a). At first sight it seems as though the use of one-point integration for the shearing terms will merely produce incorrect matrices $(\mathbf{K}_{ij})_\mathrm{s}$. However, it is easy to show that reduced integration based on the single point $x = 0$ produces an element with precisely the right bending stiffness.

The reasoning follows that set out in section 5.5. The application of one-point Gauss integration with respect to x replaces the shear term

$$b \int_{-d/2}^{d/2} \left[\int_{-L/2}^{L/2} \gamma^*(x, z)\, \tau(x, z)\, \mathrm{d}x \right] \mathrm{d}z$$

in the virtual work integral (3.7) by the expression

$$bL \int_{-d/2}^{d/2} \gamma^*(0, z)\, \tau(0, z)\, \mathrm{d}z \tag{7.8}$$

In the set of virtual strains derived from the deformation mode shown in Fig. 7.3 the virtual shear strain $\gamma^*(0, z)$ is zero for all values of z. Hence the integral (7.8) is zero for this mode of deformation. Thus reduced integration removes the contribution of the shear strains to the stiffness of the element as far as deformation in the 'bending' mode of Fig. 7.3 is concerned, which is exactly what is required. The stiffness in uniform shear is not affected. The modified matrices $(\mathbf{K}_{ij})_s$ obtained by reduced integration are

$$(\mathbf{K}_{ij})_s = (Gbd/L) \begin{bmatrix} \begin{bmatrix} 1 & L/2 \\ L/2 & L^2/4 \end{bmatrix} & \begin{bmatrix} -1 & L/2 \\ -L/2 & L^2/4 \end{bmatrix} \\ \begin{bmatrix} -1 & -L/2 \\ L/2 & L^2/4 \end{bmatrix} & \begin{bmatrix} 1 & -L/2 \\ -L/2 & L^2/4 \end{bmatrix} \end{bmatrix} \qquad (7.7b)$$

It is easy to verify that the matrix in (7.7b) has rank 1, which implies that reduced integration gives the element an additional 'degree of freedom' in shear. The 'bending' stiffness for the mode shown in Fig. 7.3 is given by equations (7.6) and (7.7a) as $m/\theta = Ebd^3/6L + GbdL/6$ and by equations (7.6) and (7.7b) as $m/\theta = Ebd^3/6L$, the latter result being in agreement with that obtained by simple (i.e. thin) beam theory.

The element based on the stiffness matrices given in (7.6) and (7.7b) is remarkably accurate and may be used with values of L/d up to about 10^4. With higher values numerical problems may be encountered, due to the fact that the individual coefficients of $(\mathbf{K}_{ij})_s$ become so much larger than those of $(\mathbf{K}_{ij})_b$ that the latter become of round-off order in the computation.

The analysis given above assumes the beam to be in a state of *plane stress* (i.e. $\sigma_{yy} = 0$). Replacing E in the bending terms by $E' = E/(1 - v^2)$ gives a *plane strain* element, which can be used to represent a strip of plate in situations where the deformation is not a function of the third coordinate y. (The shearing terms are left unchanged.) It is also straightforward to derive an element in which all loads and deformations vary as $\sin ky$, $\cos ky$ or indeed as the components of any appropriate set of orthogonal functions. A series of such elements can be used to analyse a general two-dimensional plate by the method given in section 6.5.

The analysis may also be modified to generate an axisymmetric ring element. The changes required are very similar to those described in chapter 6 for continuum elements. Again reduced integration with respect to the radial coordinate has the effect of removing the unwanted shear stiffness.

7.2   A beam element based on simple plane bending theory

An essential feature of the element described in the previous section is that the transverse displacement u of a point on the neutral axis and the rotation θ of the corresponding cross-section are regarded as *independent* variables, each one being obtained by linear interpolation from the independent nodal variables u_i and θ_i. In simple bending theory (sometimes referred to as the Euler–Bernoulli bending theory) shear strains are ignored, so that θ is equal to du/dx and the direct strain at any point is a function of the transverse displacement of the neutral axis only.

It is still legitimate to express this displacement in terms of the transverse displacements u_i and rotations θ_i ($= (du/dx)_i$) associated with nodes at each end of the element. However, these four quantities can now be combined to define a *cubic* approximation for u, in place of the *linear* approximation used in the deep beam element. The cubic approximation is shown in Fig. 7.4 and can be written in the form

$$u(x) = u_i n_i(x) + \theta_i n_{xi}(x) \quad (i = 1, 2) \tag{7.9}$$

where the one-dimensional cubic shape functions n_i and n_{xi} are the Hermite polynomials obtained as the solution to problem 1.2, with a change of variable to allow for the change to the integration range $-L/2 \leqslant x \leqslant L/2$ shown in Fig. 7.4. These polynomials are

$$\left.\begin{aligned}
n_1 &= (L^3 - 3xL^2 + 4x^3)/2L^3 \\
n_{x1} &= (L^3 - 2xL^2 - 4x^2L + 8x^3)/8L^2 \\
n_2 &= (L^3 + 3xL^2 - 4x^3)/2L^3 \\
n_{x2} &= (-L^3 - 2xL^2 + 4x^2L + 8x^3)/8L^2
\end{aligned}\right\} \tag{7.10}$$

The analysis now follows a similar pattern to that given in chapter 3, the strain 'vector' ε being replaced by the curvature κ and the stress 'vector' σ being replaced by the bending moment m, where $m = EI\kappa$, I being the second moment of area of the cross-section. The relationship between κ and the nodal variables is

$$\kappa = d^2u/dx^2 = (d^2n_i/dx^2)u_i + d^2n_{xi}/dx^2)\theta_i$$

Fig. 7.4. The deformation mode for a cubic beam element based on simple bending theory.

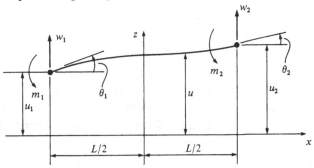

which may be written in the form

$$\kappa = [\mathrm{d}^2 n_i/\mathrm{d}x^2\, \mathrm{d}^2 n_{xi}/\mathrm{d}x^2]\, \mathbf{u}_i = \mathbf{B}_i \mathbf{u}_i$$

Substitution of the shape functions (7.10) into (7.11) gives

$$\mathbf{B}_1 = [12x/L^3\,(6x-L)/L^2], \quad \mathbf{B}_2 = [-12x/L^3\,(6x+L)/L^2]$$

For a thin beam the virtual work equation (3.7) becomes

$$\int_{-L/2}^{L/2} \kappa^* m\, \mathrm{d}x = \int_{-L/2}^{L/2} u^* w\, \mathrm{d}x$$

where w is the distribution of transverse loading. The stiffness matrices $\mathbf{K}_{ij}$ take the form

$$\mathbf{K}_{ij} = \int_{-L/2}^{L/2} EI\, \mathbf{B}_i^t \mathbf{B}_j\, \mathrm{d}x \quad (i,j = 1, 2) \tag{7.12}$$

while the equivalent nodal loads are

$$\mathbf{w}_i = \int_{-L/2}^{L/2} \mathbf{n}_i w\, \mathrm{d}x \quad (i = 1, 2) \tag{7.13}$$

where $\mathbf{w}_i = \begin{bmatrix} w_i \\ m_i \end{bmatrix}$ and $\mathbf{n}_i = \begin{bmatrix} n_i \\ n_{xi} \end{bmatrix}$. Note that $\mathbf{B}_1$ and $\mathbf{B}_2$ are *row* vectors, so that the products $\mathbf{B}_i^t \mathbf{B}_j$ are 2×2 matrices, not scalars. In contrast to the deep-beam element developed in section 7.1 the moments m_i are, in general, non-zero.

If the beam is of constant cross-section the integrand of (7.12) is quadratic in x and the matrices $\mathbf{K}_{ij}$ may be evaluated exactly using the Gauss two-point formula

$$\int_{-L/2}^{L/2} f(x)\, \mathrm{d}x \approx (L/2)\,[f(L/2\sqrt{3}) + f(-L/2\sqrt{3})]$$

to give

$$\mathbf{K}_{11} = (EI/L^3)\begin{bmatrix} 12 & 6L \\ 6L & 4L^2 \end{bmatrix}$$

$$\mathbf{K}_{12} = \mathbf{K}_{21}^t = (EI/L^3)\begin{bmatrix} -12 & 6L \\ -6L & 2L^2 \end{bmatrix} \tag{7.14}$$

$$\mathbf{K}_{22} = (EI/L^3)\begin{bmatrix} 12 & -6L \\ -6L & 4L^2 \end{bmatrix}$$

The expressions (7.14) are exact for a straight beam of constant cross-section in plane bending. They can be obtained by a number of other methods, including direct integration of the differential equations of bending. Their derivation here from an assumed deformation function bears out the statement, made in chapter 1, that the Ritz procedure will always find the exact solution of a differential equation if that solution is included in the chosen approximating function.

As with the deep-beam element, the analysis developed in this section can be modified to provide conforming strip and axisymmetric ring elements for use in appropriate thin-plate analyses. In the case of an axisymmetric constant-thickness element the use of a cubic displacement approximation produces small errors in the matrices $\mathbf{K}_{ij}$, since the exact analytic solution for the transverse displacement includes a logarithmic term as well as powers of r.

7.3   Structures formed from plane beam elements: coordinate transformation matrices

A number of thick- or thin-beam elements can be connected end-to-end to form a continuous beam. If the beam is straight the element coordinate axes x, y, z shown in Fig. 7.1 are in the same direction for all the elements and form a natural global coordinate system for the analysis. In such cases the process of connecting two elements merely involves equating the values of u_i and θ_i at the common node. In the case of thin-beam elements this process automatically ensures that the transverse displacement of the neutral axis has continuous value and derivative. As stated in section 1.3, this degree of continuity is necessary to ensure convergence of the Ritz process.

If, however, the elements form part of a curved beam, so that initially their neutral axes are not all in the same direction, the nodal loads and displacements associated with the elements must be expressed in a common global coordinate system before the assembly process is carried out. Moreover, axial forces and displacements must be included in the individual element load/displacement equations, since an axial force in one element may generate shear forces in its neighbours. The following analysis deals with the case where the elements and the applied loads all lie in a plane.

A typical element is shown in Fig. 7.5. In this element the transverse nodal displacements and shear forces (written as u_i and w_i in sections 7.1 and 7.2) are denoted by u_{zi} and w_{zi}, the axial displacements and forces being denoted by u_{xi} and w_{xi}. The relationship between axial displacements and forces is

$$(k_x)_{ij} u_{xj} = w_{xi} \quad (i,j = 1, 2) \tag{7.15}$$

where $(k_x)_{11} = (k_x)_{22} = EA/L, \quad (k_x)_{12} = (k_x)_{21} = -EA/L$.

Equations (7.15) may be combined with the nodal load/displacement equations set up in sections 7.1 or 7.2, the variables u_{xi} and w_{xi} being added to the nodal variables $\mathbf{u}_i$ and $\mathbf{w}_i$ defined in those sections and the elements $(k_x)_{ij}$ being added to the matrices $\mathbf{K}_{ij}$ appearing in equations (7.6), (7.7a), (7.7b) or (7.14). The combined equations may be written in the form

$$
\begin{bmatrix} (k_x)_{ij} & 0 & 0 \\ 0 & & \\ 0 & & \mathbf{K}_{ij} \end{bmatrix} \begin{bmatrix} u_{xj} \\ u_{zj} \\ \theta_j \end{bmatrix} = \begin{bmatrix} w_{xi} \\ w_{zi} \\ m_i \end{bmatrix}
$$

or more briefly as

$$
\mathbf{K}_{ij}\mathbf{u}_j = \mathbf{w}_i \tag{7.16}
$$

the symbols $\mathbf{K}$, $\mathbf{u}$ and $\mathbf{w}$ being re-defined appropriately. In the x, z coordinate system the 'axial' and 'bending' equations are uncoupled, but coupling occurs when the change to the global coordinate system x', z' is made.

The nodal displacement variables may be expressed in either the local x, z coordinate system or the global x', z' system, the relationship between the two sets of variables being

$$
\mathbf{u}_i' = \begin{bmatrix} u_{xi}' \\ u_{zi}' \\ \theta_i' \end{bmatrix} = \begin{bmatrix} \cos\psi & -\sin\psi & 0 \\ \sin\psi & \cos\psi & 0 \\ 0 & 0 & 1 \end{bmatrix} \begin{bmatrix} u_{xi} \\ u_{zi} \\ \theta_i \end{bmatrix} = \mathbf{T}\mathbf{u}_i \quad (i = 1, 2) \tag{7.17}
$$

Similarly the two alternative expressions for the nodal load variables satisfy the relationship $\mathbf{w}_i' = \mathbf{T}\mathbf{w}_i$. Since $\mathbf{T}$ is orthogonal, equation (7.17) may be written as $\mathbf{u}_i = \mathbf{T}^t\mathbf{u}_i'$. Substitution of these two relationships into the nodal load/displacement equations (7.16) gives

$$
\mathbf{T}\mathbf{K}_{ij}\mathbf{T}^t\mathbf{u}_j' = \mathbf{w}_i'
$$

Fig. 7.5. The change from element coordinates to global coordinates for a plane beam.

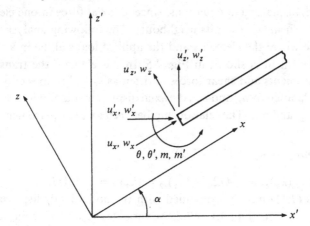

or

$$\mathbf{K}'_{ij}\mathbf{u}'_j = \mathbf{w}'_i \qquad (7.18)$$

where $\mathbf{K}'_{ij} = \mathbf{T}\,\mathbf{K}_{ij}\mathbf{T}^t$. After the nodal equations for all the elements have been converted to the form (7.18) the assembly process is carried out in the usual way.

With the above modifications the beam elements described in sections 7.1 and 7.2 can be used to analyse plane trusses, frames and polygonal arches. A curved beam can be treated approximately as a series of straight segments, the degree of approximation depending on the ratio of element length to radius of curvature. The extension of the above analysis to general three-dimensional skeletal structures is straightforward. A detailed account of the analysis of frames, trusses and arches will be found in reference 2.

7.4 Square and quadrilateral thick-plate elements

The development of nodal stiffness matrices for a square thick-plate element follows a similar pattern to that already set out for a deep beam in section 7.1. At each node three independent displacement components are defined: the transverse displacement u_i of the middle surface and the two rotations θ_{xi}, θ_{yi} of a line originally perpendicular to that surface. These variables are shown in Fig. 7.6, together with the corresponding nodal load components w_i, m_{xi}, m_{yi}. The values u, θ_x, θ_y at an arbitrary point on the middle surface are obtained from the nodal values by interpolation, using the bi-linear shape functions given in equation (4.4). As in chapter 4, it is convenient to write these functions as $n_i(\alpha, \beta)$, using

Fig. 7.6. A square thick-plate element.

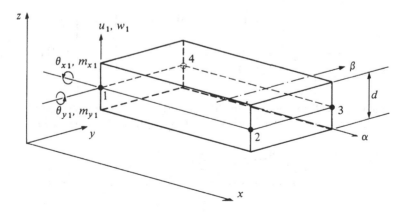

local coordinates α, β in preparation for an extension of the analysis to quadrilateral elements.

A straightforward generalisation of the assumptions (a), (b), (c) and (d) given in section 7.1 leads to the displacement approximation

$$\begin{bmatrix} u_x \\ u_y \\ u_z \end{bmatrix} = \begin{bmatrix} -z\theta_y \\ -z\theta_x \\ u \end{bmatrix} = \begin{bmatrix} -z\,n_i\,\theta_{yi} \\ -z\,n_i\,\theta_{xi} \\ n_i u_i \end{bmatrix}$$

which can be written as

$$\begin{bmatrix} u_x \\ u_y \\ u_z \end{bmatrix} = \begin{bmatrix} 0 & 0 & -zn_i \\ 0 & -zn_i & 0 \\ n_i & 0 & 0 \end{bmatrix} \begin{bmatrix} u_i \\ \theta_{xi} \\ \theta_{yi} \end{bmatrix} \tag{7.19}$$

Since u_x, u_y and u_z each vary linearly on the boundary of the element the displacement compatibility possessed by the parent eight-node brick element is maintained – i.e. the element conforms.

The relevant strains are given by equation (3.15a), omitting ε_{zz}, which is assumed to be small compared with u. Combining that equation with (7.19) gives

$$\begin{bmatrix} \varepsilon_{xx} \\ \varepsilon_{yy} \\ \gamma_{xy} \\ \gamma_{yz} \\ \gamma_{zx} \end{bmatrix} = \begin{bmatrix} 0 & 0 & -z\,\partial n_i/\partial x \\ 0 & -z\,\partial n_i/\partial y & 0 \\ 0 & -z\,\partial n_i/\partial x & -z\,\partial n_i/\partial y \\ \partial n_i/\partial y & -n_i & 0 \\ \partial n_i/\partial x & 0 & -n_i \end{bmatrix} \begin{bmatrix} u_i \\ \theta_{xi} \\ \theta_{yi} \end{bmatrix} \tag{7.20}$$

Equation (7.20) may be written in the familiar form $\varepsilon = \mathbf{B}_i \mathbf{u}_i$. Since $\partial/\partial x \equiv \partial/\partial \alpha$ and $\partial/\partial y \equiv \partial/\partial \beta$ the components of the matrices $\mathbf{B}_i$ are easily obtained from the shape functions $n_i(\alpha, \beta)$ given in equation (4.4). The stress/strain equations can be written in a form analogous to that used for the deep-beam element

$$\boldsymbol{\sigma} = \begin{bmatrix} \sigma_{xx} \\ \sigma_{yy} \\ \tau_{xy} \\ \tau_{yz} \\ \tau_{zx} \end{bmatrix} = \begin{bmatrix} E' & vE' & 0 & 0 & 0 \\ vE' & E' & 0 & 0 & 0 \\ 0 & 0 & G & 0 & 0 \\ 0 & 0 & 0 & G & 0 \\ 0 & 0 & 0 & 0 & G \end{bmatrix} \begin{bmatrix} \varepsilon_{xx} \\ \varepsilon_{yy} \\ \gamma_{xy} \\ \gamma_{yz} \\ \gamma_{zx} \end{bmatrix} = \mathbf{D}\,\boldsymbol{\varepsilon} \tag{7.21}$$

where $E' = E/(1-v^2)$. As in the deep beam, σ_{zz} is assumed to be zero.

The stiffness matrices $\mathbf{K}_{ij}$ are 3×3 matrices and take the usual form

$$\mathbf{K}_{ij} = \iiint_V \mathbf{B}_i^t \mathbf{D} \mathbf{B}_j \, d\alpha \, d\beta \, dz \tag{7.22}$$

The equivalent nodal loads produced by a transverse distribution of load

$w(x, y)$ are given by

$$w_i = \iint_A n_i w \, d\alpha \, d\beta, \quad m_{xi} = m_{yi} = 0 \qquad (7.23)$$

As with the deep beam, the integrand in (7.22) may be split up into 'bending' terms, multiplied by E', and 'shearing' terms, multiplied by G. Since the matrices $\mathbf{B}_i$ are linear in α and β, both the bending and the shearing contributions to $\mathbf{K}_{ij}$ can be evaluated exactly by using a four-point Gauss formula for integration with respect to α and β. However, the performance of the element is greatly improved if one-point integration based on the origin of the local coordinates α, β is used for the shearing terms. The reason for this improvement is exactly the same as that given in section 7.1. Reduced integration removes the spurious contribution from the shear strains associated with the 'bending' deformation mode of Fig. 7.3, without affecting the stiffness of the element with respect to either uniform shear or twisting.

As in plane-stress problems, the bi-linear shape functions $n_i(\alpha, \beta)$ which appear in equation (7.19) may be used as mapping functions, generating the general quadrilateral element shown in Fig. 7.7. The procedure is identical to that given in chapter 5. The components of the nodal rotations θ_i must be expressed in the global x, y coordinate system, as shown in Fig. 7.7a, since equality of θ at a node where several elements meet requires the components of $\boldsymbol{\theta}$ to be in a coordinate system common to all the elements.

The simple four-node element described in this section gives accurate

Fig. 7.7. Mapping the square element to give a quadrilateral thick-plate element.

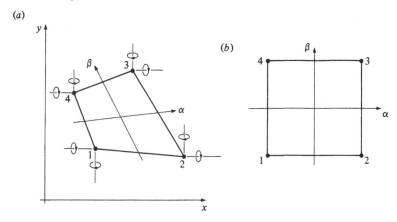

results for values of L/d up to about 10^5. Higher-order elements may also be developed along the lines set out in chapters 4 and 5, curved boundaries being treated by the use of non-linear mapping functions in the manner described for plane-stress elements.

7.5   A square thin-plate element: the problem of boundary continuity

The change from a square thick-plate element to a square thin-plate element follows exactly the same pattern as the change from a deep-beam element to a thin-beam element. As with the thin beam, the assumption that transverse shear strains can be ignored implies that the strain at any point in a thin plate is a function of the transverse displacement of the middle surface only.

The square element considered in this section is shown in Fig. 7.8. This element is similar to that shown in Fig. 7.6, except that the rotations θ_{x1}, θ_{y1}, ... are now equal to $(\partial u/\partial y)_1$, $(\partial u/\partial x)_1$, ... The displacement u of the middle surface may be expressed in terms of the twelve nodal variables u_i, θ_{xi}, θ_{yi} as

$$u(\alpha, \beta) = u_i n_i + \theta_{xi} n_{yi} + \theta_{yi} n_{xi} \quad (i = 1, ..., 4) \tag{7.24}$$

The functions n_i, n_{xi}, n_{yi} are the two-dimensional shape functions

$$\left.\begin{aligned}
n_i &= (1 + \alpha_i \alpha)(1 + \beta_i \beta)(2 + \alpha_i \alpha + \beta_i \beta - \alpha^2 - \beta^2)/8 \\
n_{xi} &= (\alpha^2 - 1)(\alpha + \alpha_i)(1 + \beta_i \beta)/8 \\
n_{yi} &= (\beta^2 - 1)(\beta + \beta_i)(1 + \alpha_i \alpha)/8
\end{aligned}\right\} \tag{7.25}$$

where α, β are the local coordinates shown in Fig. 7.8 and α_i, β_i are the vertices of the element, i.e. the points $(\pm 1, \pm 1)$. Equations (7.24) and (7.25) are simply the two-dimensional versions of equations (7.9) and

Fig. 7.8. A square thin-plate element.

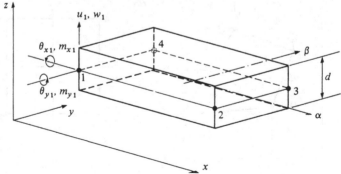

(7.10). It is easy to verify that the shape functions defined in (7.25) have the required properties. Their general form is shown in Fig. 7.9.

The change from thin beam to thin plate results in the scalar curvature κ introduced in equation (7.11) being replaced by a vector $\boldsymbol{\kappa}$ with components $\partial^2 u/\partial x^2, \partial^2 u/\partial y^2, 2\partial^2 u/\partial x\partial y$.† The scalar bending moment m is similarly replaced by a vector $\mathbf{m}$ with components m_{xx}, m_{yy}, m_{xy}. The six quantities which make up $\boldsymbol{\kappa}$ and $\mathbf{m}$ are shown in Fig. 7.10. They are

† The factor 2 appears because the term in the strain energy expression associated with the twisting moment m_{xy} is $2m_{xy}\partial^2 u/\partial x\,\partial y$.

Fig. 7.9. (*a*) The cubic shape function $n_3(\alpha, \beta)$.
(*b*) The cubic shape function $n_{x3}(\alpha, \beta)$.

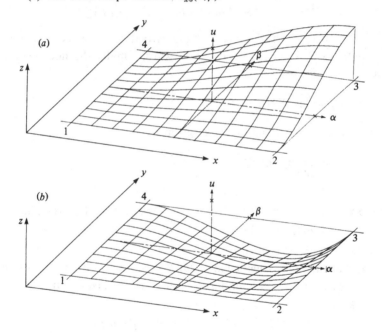

Fig. 7.10. Sign conventions for bending moments m_{xx}, m_{yy}, curvatures κ_{xx}, κ_{yy}, twisting moment m_{xy} and torsion κ_{xy}.

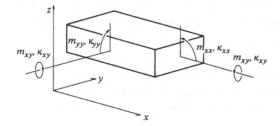

related by the constitutive equation

$$
\mathbf{m} = \begin{bmatrix} m_{xx} \\ m_{yy} \\ m_{xy} \end{bmatrix} = \frac{Ed^3}{12(1-v^2)} \begin{bmatrix} 1 & v & 0 \\ v & 1 & 0 \\ 0 & 0 & (1-v)/2 \end{bmatrix} \begin{bmatrix} \partial^2 u/\partial x^2 \\ \partial^2 u/\partial y^2 \\ 2\partial^2 u/\partial x\,\partial y \end{bmatrix} = \mathbf{D}\boldsymbol{\kappa} \quad (7.26)
$$

The relationship between κ and the nodal variables $\mathbf{u}_i$ is

$$
\boldsymbol{\kappa} = \begin{bmatrix} \partial^2 u/\partial x^2 \\ \partial^2 u/\partial y^2 \\ 2\partial^2 u/\partial x\,\partial y \end{bmatrix}
$$

$$
= \begin{bmatrix} \partial^2 n_i/\partial x^2 & \partial^2 n_{yi}/\partial x^2 & \partial^2 n_{xi}/\partial x^2 \\ \partial^2 n_i/\partial y^2 & \partial^2 n_{yi}/\partial y^2 & \partial^2 n_{xi}/\partial y^2 \\ 2\partial^2 n_i/\partial x\,\partial y & 2\partial^2 n_{yi}/\partial x\,\partial y & 2\partial^2 n_{xi}/\partial x\,\partial y \end{bmatrix} \begin{bmatrix} u_i \\ \theta_{xi} \\ \theta_{yi} \end{bmatrix} = \mathbf{B}_i \mathbf{u}_i \quad (7.27)
$$

Since $\partial/\partial x \equiv \partial/\partial \alpha$ and $\partial/\partial y \equiv \partial/\partial \beta$ the derivation of the components of the matrices $\mathbf{B}_i$ in (7.27) is straightforward. Finally the nodal stiffness matrices are given by

$$
\mathbf{K}_{ij} = \iint_A \mathbf{B}_i^t \mathbf{D}\mathbf{B}_j \, d\alpha \, d\beta \quad (7.28)
$$

and the equivalent nodal loads are given by

$$
\mathbf{w}_i = \iint_A \mathbf{n}_i w \, d\alpha \, d\beta \quad (7.29)
$$

where $\mathbf{w}_i = \begin{bmatrix} w_i \\ m_{xi} \\ m_{yi} \end{bmatrix}$, $\mathbf{n}_i = \begin{bmatrix} n_i \\ n_{yi} \\ n_{xi} \end{bmatrix}$, w is the transverse load on the plate and

A is the area of the element.

The shape functions defined in (7.25) vary cubically with α and β on each edge of the element. On boundary 2–3, for example, equations (7.24) and (7.25) give

$$
u = [(1-\beta)(2-\beta-\beta^2)u_2 + (1+\beta)(2+\beta-\beta^2)u_3
$$
$$
+ (\beta^2-1)(\beta-1)\theta_{x2} + (\beta^2-1)(\beta+1)\theta_{x3}]/4
$$

This is precisely the one-dimensional cubic interpolating polynomial given by equations (7.9) and (7.10). The four nodal values $u_2, u_3, \theta_{x2}, \theta_{x3}$ define a unique cubic displacement on the edge 2–3, ensuring continuity of u if this edge is a common boundary between two elements.

Although the deformation defined by (7.24) and (7.25) makes u continuous in *value* across an inter-element boundary it does not, in general, give continuity of *normal derivative*, except at the nodes. For example, consider the two elements A and B shown in Fig. 7.11. Let all the nodal variables associated with these elements be zero except for θ_{y3} in element *B*, which is equal to 1. Within *A*, *u* is zero, so that $(\partial u/\partial y)_A = 0$ on the boundary p, ..., q. Within *B*, *u* is equal to $(\alpha^2 - 1)(1 + \alpha)(1 + \beta)/8$, so that $(\partial u/\partial y)_N = (\alpha^2 - 1)(1 + \alpha)/8$ on the boundary p, ..., q. The latter expression for $\partial u/\partial y$ represents the discontinuity of normal derivative on the boundary, indicating that the element does not conform.

This lack of continuity of derivative does not necessarily make the element useless and indeed it is capable of giving satisfactory accuracy in practical problems. It is straightforward to verify that if nodal values corresponding to any condition of constant curvature or torsion are substituted into (7.24) then equation (7.27) gives the correct (constant) value of κ throughout the element, implying continuity of both displacement and normal derivative on all inter-element boundaries under these conditions. This is essentially the same condition as that established in section 4.6 (equations (4.9)), and implies that the element satisfies the patch test described in section 5.6. As stated in that section, satisfaction of the patch test guarantees convergence to the correct solution as the element size is reduced.

The square element is easily generalised to a rectangle with very similar convergence properties. It is posible to map the element into a general quadrilateral using the bi-linear shape functions defined in equation (4.4). However, the mapping process produces an element which does not satisfy the constant-curvature condition. Convergence is therefore no longer guaranteed and indeed the performance of the quadrilateral tends to be unsatisfactory.

It is also possible to produce triangular thin-plate elements with three variables at each node – one displacement and two gradient components.

Fig. 7.11. Example showing lack of continuity of normal derivative on the boundary between two four-node plate elements.

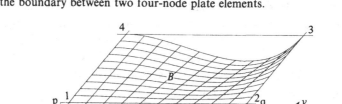

Such elements do not conform, but the equilateral triangle satisfies the patch test. Less regular triangles do not, in general, satisfy this test and do not, therefore, always give convergence to the correct solution. However, the errors are usually small and the elements are often used in practice.

The search for more accurate elements is largely a search for elements which give continuity of normal derivative at all points on the inter-element boundaries. It is impossible to produce such elements using polynomials based on nodal values of u and its first derivatives, and a number of alternative approaches have been suggested, including:

(a) The introduction of $(\partial^2 u/\partial x\, \partial y)_i$ as a fourth variable at each node. This allows the development of conforming square and rectangular elements, but difficulties arise if the procedure is used with more general quadrilateral elements.

(b) The introduction of mid-side nodes at which only derivative variables are defined. This may give an improvement in accuracy, but the fact that there are different numbers of variables associated with different nodes makes the organisation of the assembly process more complex.

(c) The introduction of shape functions which have discontinuous second derivatives.

These ideas are discussed in detail in reference 1.

7.6   Structures formed from plate elements

Just as a number of beam elements can be connected to form a straight beam, so a number of plate elements can be connected to form a flat plate. If square or rectangular elements are used, the coordinates x, y, z shown in Fig. 7.6 or Fig. 7.8 are the same for all elements and form a global coordinate system for the whole plate. In such cases the assembly of the nodal load/displacement equations is straightforward.

Plate elements can also be connected in such a way that the middle surfaces of the various elements do not all lie in the same plane. A box-girder bridge is an obvious practical example. The assembly of the nodal equations for such structures follows a pattern similar to that set out for beam elements in section 7.3.

The analysis begins with the addition of plane-stress degrees of freedom to the individual plate-bending elements. The nodal displacement variables u_i (now written as u_{zi}), θ_{xi}, θ_{yi} introduced in sections 7.4 and 7.5 are combined with in-plane displacement components u_{xi} and u_{yi}, with corresponding changes in the nodal load variables. The plate-bending stiffness matrices defined by equations (7.22) or (7.28) are combined with suitable plane-stress stiffness matrices, as derived in chapters 3, 4 and 5. It is common practice also to add a rotational degree of freedom θ_{zi},† giving six degrees of freedom at each node – three displacements and three rotations. The complete set of nodal equations is written in the standard form

$$\mathbf{K}_{ij}\mathbf{u}_j = \mathbf{w}_i \tag{7.30}$$

the components of the nodal displacements $\mathbf{u}_j$ being arranged in the order $u_{xj}, u_{yj}, u_{zj}, \theta_{xj}, \theta_{yj}, \theta_{zj}$.

The next step is the transformation of equations (7.30) into the global coordinate system x', y', z' shown in Fig. 7.12. The analysis is simply the three-dimensional version of that given in section 7.3. The transformed equations are

$$\mathbf{K}'_{ij}\mathbf{u}'_j = \mathbf{w}'_i \tag{7.31}$$

where $\mathbf{K}'_{ij} = \mathbf{T}\,\mathbf{K}_{ij}\,\mathbf{T}^t$. The transformation matrix $\mathbf{T}$ is given by

$$\mathbf{T} = \begin{bmatrix} xx' & xy' & xz' & 0 & 0 & 0 \\ yx' & yy' & yz' & 0 & 0 & 0 \\ zx' & zy' & zz' & 0 & 0 & 0 \\ 0 & 0 & 0 & xx' & xy' & xz' \\ 0 & 0 & 0 & yx' & yy' & yz' \\ 0 & 0 & 0 & zx' & zy' & zz' \end{bmatrix} \tag{7.32}$$

where xx', xy', etc. indicate the cosines of the angles between the specified *element* and *global* axes. After the nodal equations for all the elements have

† No stiffness or external load is associated with this degree of freedom.

Fig. 7.12. The change from element coordinates to global coordinates for a flat plate.

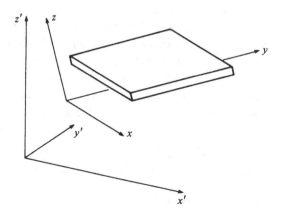

been converted to the form (7.31) the assembly process is carried out in the usual way.

It is important to realise that the expression of the nodal variables in two alternative coordinate systems x, y, z and x', y', z' is logically quite distinct from the change to parametric coordinates α, β described in chapter 5. The introduction of parametric coordinates is merely a device which simplifies the calculation of the stiffness matrices $\mathbf{K}_{ij}$, and is done *after* the nodal variables have been defined.

With the above modifications the plate elements described in sections 7.4 and 7.5 can be used to analyse a box girder or similar 'folded-plate' structure. They can also be used to give an approximate representation of a curved shell, the shell being replaced by a series of flat plates. Shells with single curvature, such as tubes and barrel-vaulted roofs, can be modelled using square or rectangular plate elements, while shells with double curvature require triangular elements.

7.7   Shell elements

When a shell is represented by an assembly of flat plates the coupling between 'bending' and 'stretching' deformations takes place only on the boundaries between the plates. In a finite-element analysis of a shell structure which uses flat-plate elements the coupling is produced by the transformation of the element nodal stiffness matrices into a common global coordinate system, as described in section 7.6.

By contrast, in a segment of shell which has a curved middle-surface the coupling takes place within each infinitesimal portion of the segment. This is the reason why the governing differential equation of a shell is considerably more complicated than that of a plate. It is also the reason why the development of shell finite elements has proceeded at a much slower pace than the development of plate elements.

The simplest shell element is the conical axisymmetric ring element shown in Fig. 7.13. If this element is treated as a 'thin' shell then the analysis follows a very similar pattern to that set out for a thin beam in section 7.2. The displacement vector $\mathbf{u}$ associated with the point X[†] on the neutral axis in Fig. 7.13a has two components u_x and u_z. The displacement component u_x is expressed in terms of the nodal displacements u_{xi} using linear interpolation, i.e.

$$u_x = (\tfrac{1}{2}-x/L)u_{x1}+(\tfrac{1}{2}+x/L)u_{x2} \tag{7.33a}$$

while the displacement component u_z is expressed in terms of the nodal

† Strictly a circle of points on the neutral surface.

displacements u_{zi} and rotations θ_i using the cubic interpolation equation (7.9),

$$u_z = u_{zi}n_i + \theta_i n_{xi} \quad (i = 1, 2) \tag{7.33b}$$

the polynomials n_i and n_{xi} being given by equation (7.10).

In thin-shell theory the 'strain' vector ε consists of the extensional strains ε_{xx}, $\varepsilon_{\phi\phi}$ and the curvature *changes* κ_{xx}, $\kappa_{\phi\phi}$. These are related to the displacement **u** by the equations

$$\varepsilon = \begin{bmatrix} \varepsilon_{xx} \\ \varepsilon_{\phi\phi} \\ \kappa_{xx} \\ \kappa_{\phi\phi} \end{bmatrix} = \begin{bmatrix} \mathrm{d}u_x/\mathrm{d}x \\ (u_z \cos\alpha + u_x \sin\alpha)/r \\ \mathrm{d}^2 u_z/\mathrm{d}x^2 \\ (\sin\alpha/r)\,\mathrm{d}u_z/\mathrm{d}x \end{bmatrix} \tag{7.34}$$

Substitution of expressions (7.33a) and (7.33b) into (7.34) gives an

Fig. 7.13. A conical axisymmetric shell element.
(*a*) Notation and sign conventions for nodal variables.
(*b*) Notation and sign conventions for internal moments and forces.

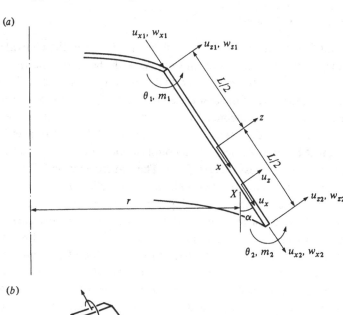

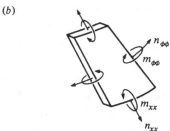

expression for ε in terms of the nodal variables which can be written in the usual way as $\varepsilon = \mathbf{B}_i \mathbf{u}_i$, each matrix $\mathbf{B}_i$ being a 4×2 matrix whose elements are functions of x. (Note that r is a linear function of x.) The 'stress' vector $\boldsymbol{\sigma}$ which corresponds to ε consists of the quantities n_{xx}, $n_{\phi\phi}$ (tangential force per unit length) and m_{xx}, $m_{\phi\phi}$ (moments per unit length) shown in Fig. 7.13b. The constitutive matrix relating $\boldsymbol{\sigma}$ and ε is

$$\mathbf{D} = \frac{Ed}{1-v^2} \begin{bmatrix} 1 & v & 0 & 0 \\ v & 1 & 0 & 0 \\ 0 & 0 & d^2/12 & vd^2/12 \\ 0 & 0 & vd^2/12 & d^2/12 \end{bmatrix} \tag{7.35}$$

Finally the element nodal stiffness matrices are given by

$$\mathbf{K}_{ij} = \int_A \mathbf{B}_i^t \mathbf{D} \mathbf{B}_j \, dA \quad (i,j = 1,2)$$

where integration is over the conical middle-surface of the element. Integrating with respect to ϕ gives

$$\mathbf{K}_{ij} = 2\pi \int_{-L/2}^{L/2} \mathbf{B}_i^t \mathbf{D} \mathbf{B}_j r \, dx$$

If external forces act on the surface of the element then the equivalent nodal loads are obtained in the usual way. As with the axisymmetric ring elements described in chapter 6, each nodal load component is distributed round the circumference of the associated nodal circle. If a doubly-curved axisymmetric shell is represented by a number of conical elements with different values of apex angle 2α then the nodal load/displacement equations must be transformed into a common global coordinate system before the assembly process is carried out. The transformation follows the same pattern as that given for beam elements in section 7.3.

The conical element developed in this section can be generalised (a) by making the centre-line of the element curved in the x, z plane and (b) by making the displacement and loading vary as $\cos \phi$ or $\sin \phi$, as discussed in section 6.5. Full details of these extensions will be found in reference 1.

Thick-shell axisymmetric elements can also be constructed, the analysis being similar to that already given for a thick beam. As with the thick-beam and thick-plate elements described in sections 7.1 and 7.4, the use of reduced integration for the shear stiffness produces a considerable improvement in accuracy.

Finally, a general doubly-curved thick-shell element can be obtained by first distorting a suitable three-dimensional element, such as the 16-node square brick element shown in Fig. 7.14, by means of the mapping technique described in chapter 5. The sixteen nodes with three degrees of freedom at each node are then replaced by eight nodes on the middle surface, the procedure being similar to that described in sections 7.1 and 7.4. At each node of the modified element five degrees of freedom are defined – two tangential 'stretching' displacements, one transverse displacement and two rotations about axes tangential to the middle surface. The analysis appears more complicated than that set out in section 7.4, but this complication is largely due to the need to use curvilinear coordinates based on the curved middle surface of the element.

Problems for chapter 7

7.1 Obtain approximate values for the tip deflection of the cantilever considered in section 4.9, representing the cantilever by

(*a*) a single deep-beam element, as developed in section 7.1, using (i) exact integration (equation (7.7*a*)) and (ii) reduced integration (equation 7.7*b*)) for the evaluation of the stiffness matrices, and

(*b*) a single thin-beam element, as described in section 7.2.

Compare your results with the exact solution given in equation (4.19).

7.2 Make a similar comparison for the case $L/d = 10$.

Fig. 7.14. A 16-node brick element transformed into a general curvilinear shell element.

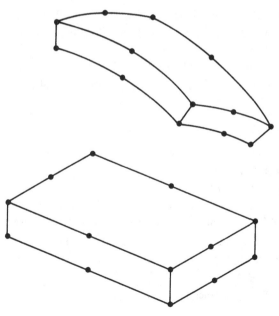

7.3 Use the shape functions n_i, n_{zi} defined in equation (7.10) to find the equivalent nodal loads for the thin-beam element developed in section 7.2 due to

(a) a uniform transverse load of w per unit length, and

(b) a point load W at a distance a from one end.

7.4 Find the stiffness matrices and equivalent nodal loads for the axisymmetric annular thick-plate element obtained by generalising the deep-beam element developed in section 7.1.

7.5 Generalise the thin-beam element developed in section 7.2 to give the stiffness matrices and equivalent nodal loads for a thin-plate element of width L and length b, assuming the transverse displacement $u(x, y)$ to be

of the form $\sum_k u^{(k)}(x) \sin(k\pi y/b)$.

7.6 Verify that the square thin-plate element developed in section 7.5 reproduces the exact solution, (with continuity of displacement and normal derivative on its boundaries), when the nodal displacements and slopes are those associated with conditions of constant curvature or torsion.

Solutions to problems

7.1 The exact tip deflection is given by equation (4.19). Substitution of $I = d^3/12$, $L/d = 2$ and $v = 0.3$ into that equation gives a deflection $u = -37.5 \, W/E$.

(a) (i) For a cantilever built-in at node 2 only K_{11} is relevant. Using equations (7.6) and (7.7a) with $b = 1$ gives the load/displacement equations for node 1

$$\begin{bmatrix} -W \\ 0 \end{bmatrix} = \begin{bmatrix} Gd/L & Gd/2 \\ Gd/2 & GdL/3 + Ed^3/12L \end{bmatrix} \begin{bmatrix} u \\ \theta \end{bmatrix}$$

Elimination of the rotation θ gives

$$u = -\frac{WL}{Gd} \frac{4G + Ed^2/L^2}{G + Ed^2/L^2}$$

Substituting $G = (\frac{5}{6}) E/2(1+v)$, $L/d = 2$, $v = 0.3$ gives a deflection $u = -16.8 \, W/E$, which is 45% of the exact value given above.

(ii) Similar analysis using (7.7b) in place of (7.7a) gives

$$u = -\frac{WL}{Gd} \frac{3G + Ed^2/L^2}{Ed^2/L^2} = -30.2 \, W/E$$

which is 81% of the exact value.

(*b*) Equation (7.14) gives the load/displacement equations

$$\begin{bmatrix} -W \\ 0 \end{bmatrix} = \frac{Ed^3}{12L^3}\begin{bmatrix} 12 & 6L \\ 6L & 4L^2 \end{bmatrix}\begin{bmatrix} u \\ \theta \end{bmatrix}$$

Elimination of θ gives $u = -4WL^3/Ed^3 = -32\,W/E$, which is 85% of the exact value. (This analysis is essentially the exact solution of section 4.9 without the shear-deflection terms.)

7.2 If $L/d = 10$ the corresponding values are,

exact	$-4027\,W/E$,
deep beam	$-122\,W/E$ (3% of the exact value),
deep beam (reduced integration)	$-3031\,W/E$ (75% of the exact value),
thin beam	$-4000\,W/E$ (99% of the exact value).

7.3 (*a*) For uniform transverse loading w, equation (7.13) becomes

$$w_i = w\int_{-L/2}^{L/2} n_i\,\mathrm{d}x, \quad m_i = w\int_{-L/2}^{L/2} n_{xi}\,\mathrm{d}x \quad (i = 1, 2)$$

where n_i and n_{xi} are defined in (7.10). Since the shape functions are cubics, two-point Gauss integration is exact, giving

$$w_i = (wL/2)[n_i(L/2\sqrt{3}) + n_i(-L/2\sqrt{3})]$$
$$m_i = (wL/2)[n_{xi}(L/2\sqrt{3}) + n_{xi}(-L/2\sqrt{3})]$$

leading to the familiar results

$$w_i = wL/2, \quad m_i = \pm wL^2/12$$

(*b*) For a single point load W distant a from end 1 equation (7.13) becomes

$$w_1 = Wn_1(-L/2 + a) = W(1 - 3a^2/L^2 + 2a^3/L^3)$$
$$m_1 = Wn_{x1}(-L/2 + a) = Wa(L-a)^2/L^2$$

with similar results for w_2 and m_2.

7.4 The element is shown in Fig. 7.15. The analysis is very similar to that developed in the solution to problem 6.1, and uses the same linear shape functions $n_i = (L/2 + \alpha_i \alpha)/L$. The displacement approximation is given by (7.1) as

$$\begin{bmatrix} u_r \\ u_z \end{bmatrix} = \begin{bmatrix} 0 & -zn_i \\ n_i & 0 \end{bmatrix}\begin{bmatrix} u_i \\ \theta_i \end{bmatrix} \tag{7.36}$$

while the strains are given by (6.10*a*), omitting the axial strain component which is assumed to be zero, i.e.

$$\varepsilon = \begin{bmatrix} \varepsilon_{rr} \\ \varepsilon_{\phi\phi} \\ \gamma_{rz} \end{bmatrix} = \begin{bmatrix} \partial/\partial r & 0 \\ 1/r & 0 \\ \partial/\partial z & \partial/\partial r \end{bmatrix}\begin{bmatrix} u_r \\ u_z \end{bmatrix} \tag{7.37}$$

Combining (7.36) and (7.37) gives

$$\varepsilon = \begin{bmatrix} 0 & -z\alpha_i/L \\ 0 & -zn_i/(r_0 + \alpha) \\ \alpha_i/L & -n_i \end{bmatrix}\begin{bmatrix} u_i \\ \theta_i \end{bmatrix} = \mathbf{B}_i\mathbf{u}_i$$

Since the plate is thin, $\mathbf{D}$ takes the form

$$\mathbf{D} = \begin{bmatrix} E' & vE' & 0 \\ vE' & E' & 0 \\ 0 & 0 & G \end{bmatrix}$$

where $E' = E/(1-v^2)$.

The stiffness matrices are given by

$$\mathbf{K}_{ij} = 2\pi \int_{-d/2}^{d/2} \int_{-L/2}^{L/2} \mathbf{B}_i^t \mathbf{D} \mathbf{B}_j (r_0 + \alpha) \, d\alpha \, dz \qquad (7.38)$$

As with the deep-beam element developed in section 7.1, the integrand in (7.38) may be split up into 'bending' terms multiplied by E' and 'shearing' terms multiplied by G. The 'bending' component of $\mathbf{K}_{ij}$ is given by

$$(\mathbf{K}_{ij})_b = \begin{bmatrix} 0 & 0 \\ 0 & K_{ij} \end{bmatrix}$$

where

$$K_{ij} = \frac{2\pi E d^3}{12(1-v^2)} [\alpha_i \alpha_j r_0 / L + (1+v)(\alpha_i + \alpha_j)/2$$
$$+ \{1/4 - r_0(\alpha_i + \alpha_j)/2L + \alpha_i \alpha_j r_0^2/L^2\} \log (r_2/r_1)]$$

(Note the similarity to the solution to problem 6.1.)

The 'shearing' component may be arranged in a similar fashion to (7.7a) as

$$(\mathbf{K}_{ij})_s =$$

$$\frac{2\pi G d r_0}{L} \begin{bmatrix} \begin{bmatrix} 1 & L/2 - L^2/12 r_0 \\ L/2 - L^2/12 r_0 & L^2/3 - L^3/24 r_0 \end{bmatrix} & \begin{bmatrix} -1 & L/2 + L^2/12 r_0 \\ -L/2 + L^2/12 r_0 & L^2/6 \end{bmatrix} \\ \begin{bmatrix} -1 & -L/2 + L^2/12 r_0 \\ L/2 + L^2/12 r_0 & L^2/6 \end{bmatrix} & \begin{bmatrix} 1 & -L/2 - L^2/12 r_0 \\ -L/2 - L^2/12 r_0 & L^2/3 - L^3/24 r_0 \end{bmatrix} \end{bmatrix}$$

Fig. 7.15.

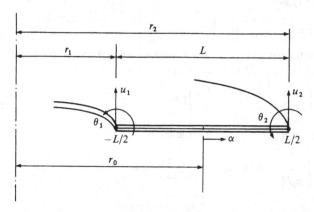

If reduced (one-point) integration is used, the expression for $(\mathbf{K}_{ij})_s$ becomes equal to (7.7b) with b replaced by $2\pi r_0$.

7.5 The harmonic component $u^{(k)}(x, y)$ of the transverse displacement $u(x, y)$ may be written in the form

$$u^{(k)}(x, y) = u^{(k)}(x) \sin (k\pi y/b) = \{n_i \sin (k\pi y/b)\} u_i^{(k)}$$
$$+ \{n_{xi} \sin (k\pi y/b)\} \theta_i^{(k)} \quad (i = 1, 2)$$

where the shape functions n_i, n_{xi} are given by (7.10). Substitution of this expression in equation (7.27) gives

$$\kappa = \begin{bmatrix} \dfrac{\partial^2 u}{\partial x^2} \\[2mm] \dfrac{\partial^2 u}{\partial y^2} \\[2mm] \dfrac{2\partial^2 u}{\partial x\,\partial y} \end{bmatrix} = \begin{bmatrix} \dfrac{\mathrm{d}^2 n_i}{\mathrm{d}x^2} \sin (k\pi y/b) & \dfrac{\mathrm{d}^2 n_{xi}}{\mathrm{d}x^2} \sin (k\pi y/b) \\[3mm] -n_i \dfrac{k^2\pi^2}{b^2} \sin (k\pi y/b) & -n_{xi} \dfrac{k^2\pi^2}{b^2} \sin (k\pi y/b) \\[3mm] \dfrac{2k\pi}{b} \dfrac{\mathrm{d}n_i}{\mathrm{d}x} \cos (k\pi y/b) & \dfrac{2k\pi}{b} \dfrac{\mathrm{d}n_{xi}}{\mathrm{d}x} \cos (k\pi y/b) \end{bmatrix} \begin{bmatrix} u_i \\[2mm] \theta_i \end{bmatrix} = \mathbf{B}_i \mathbf{u}_i$$

$$(7.39)$$

Hence

$$\mathbf{K}_{ij}^{(k)} = \int_0^b \int_{-L/2}^{L/2} (\mathbf{B}_i^{(k)})^t \, \mathbf{D}\mathbf{B}_j^{(k)} \, \mathrm{d}x \, \mathrm{d}y$$

where $\mathbf{D}$ is given by equation (7.26). As in the solution to problem 6.5, integration with respect to y gives

$$\mathbf{K}_{ij}^{(k)} = (b/2) \int_{-L/2}^{L/2} (\bar{\mathbf{B}}_i^{(k)})^t \, \mathbf{D}\bar{\mathbf{B}}_j^{(k)} \, \mathrm{d}x \qquad (7.40)$$

the matrices $\bar{\mathbf{B}}_i^{(k)}$ being given by expression (7.39) with the multipliers $\sin (k\pi y/b)$ and $\cos (k\pi y/b)$ omitted. All the terms of the matrix products in (7.40) are polynomials of degree 6 or less and the integrals may be evaluated analytically. If numerical integration is used, four-point Gauss integration is exact.

The treatment of the loading $\mathbf{w}$ follows a similar pattern, the nodal loads being given by

$$\mathbf{w}_i^{(k)} = \int_0^b \int_{-L/2}^{L/2} w^{(k)} \begin{bmatrix} n_i \\ n_{xi} \end{bmatrix} \sin^2 (k\pi y/b) \, \mathrm{d}x \, \mathrm{d}y = (b/2) \int_{-L/2}^{L/2} w^{(k)} \begin{bmatrix} n_i \\ n_{xi} \end{bmatrix} \mathrm{d}x$$

7.6 It is convenient to consider the constant-curvature surface associated with the nodal values

$$\theta_{y1} = \theta_{y4} = -1, \quad \theta_{y2} = \theta_{y3} = 1$$
$$\theta_{xi} = u_i = 0 \quad (i = 1, \dots, 4)$$

The interpolating polynomial is $u = n_{x2} - n_{x1} + n_{x3} - n_{x4}$, which on substitution from (7.25) reduces to $u = (\alpha^2 - 1)/2$, giving $\kappa = \begin{bmatrix} 1 \\ 0 \\ 0 \end{bmatrix}$ as required. A similar argument applies for curvature in the y-direction.

A convenient constant-torsion surface is the one associated with nodal values

$$u_1 = u_3 = 1, \quad u_2 = u_4 = -1$$
$$\theta_{y1} = \theta_{y2} = -1, \quad \theta_{y3} = \theta_{y4} = 1$$
$$\theta_{x1} = \theta_{x4} = -1, \quad \theta_{x2} = \theta_{x3} = 1$$

The interpolating polynomial is therefore

$$u = n_1 - n_2 + n_3 - n_4 - n_{x1} - n_{x2} + n_{x3} + n_{x4} - n_{y1} + n_{y2} + n_{y3} - n_{y4}$$

which reduces to $u = \alpha\beta$, giving $\kappa = \begin{bmatrix} 0 \\ 0 \\ 2 \end{bmatrix}$ as required.

8

Programming the finite-element method

Students learning a manual method of analysis normally build up confidence and understanding by working through examples, which are essentially special cases of a general theory. This traditional approach poses problems to those striving to become familiar with the finite-element method. It is difficult to find algebraic examples which demand more than the reproduction of standard bookwork, while numerical examples all too often merely require tedious matrix manipulations, carried out in a mechanical fashion.

This difficulty springs from the fact that the finite-element method is a systematic technique which finds its practical realisation in computer programs. Consequently a balanced appreciation of the method as a practical tool of analysis requires understanding of the task facing the programmer, rather than the ability to perform the associated calculations manually. There is no doubt that this understanding is obtained most clearly by those who have the time and opportunity to write a finite-element program for themselves.

The first part of this chapter gives a general account of the organisation of a typical finite-element program and indicates the amount of effort required in the various phases of its construction. It should be realised that this account refers primarily to the development of a 'personal' program designed to improve the understanding of the writer or provide a 'test-bed' for numerical experiments with a particular aspect of the method. It should not be taken as a guide to the construction of a large commercial finite-element program such as ANSYS, PAFEC, or NASTRAN, which is a task several orders of magnitude greater. Comments on the facilities provided by such programs will be found in the final section of the chapter.

In this chapter details of programming procedure are discussed in the context of finite-element programs for stress analysis. However, the same general arguments apply in other areas of application, with words such as 'stresses' and 'displacements' replaced by appropriate terms from the relevant discipline.

8.1   An overall strategy for program design

When the decision to write a program is taken it is desirable to prepare an 'external' specification of the program before detailed design begins. This defines the *function* of the program as it appears to the user, even though at this stage no decisions have been taken as to *how* the various operations will be carried out. In the case of a finite-element program the specification should include the following items.

(a) The type or types of problem to be solved. Although these may be stated in physical terms – heat conduction, plane stress, fluid flow, etc. the significant information from the programmer's point of view is the number of scalar unknowns associated with each node – 1 for potential theory, 2 for plane stress, 3 for plate bending, etc. It is obviously much simpler to design a program if only one type of problem is specified.

(b) Maximum values for the numbers of nodes, elements, right-hand sides, etc. These numbers control the sizes of arrays used in the program. They have a considerable effect on program organisation since in most computers fast storage is limited and there is therefore a critical size of problem above which data-packing techniques are required. Some of these techniques are discussed in sections 8.3 and 8.4.

(c) The layout of the input data. This is discussed in more detail in section 8.2.

(d) The layout of the output, including provision for graphical display of the finite-element mesh, stress contours, flow lines, etc.

This specification may well represent the wishes of an outside 'customer', rather than the decisions of the programmer, and it may alter, or require alteration, during the period of program construction. However, it is desirable that at least *tentative* decisions are taken on all the items in the above list. In particular, it is good practice to write the first draft of the program-users' manual *before* detailed coding starts, since it is this manual which defines the interface between the program and the outside world.

Once the program specification has been laid down, estimates can be made of resource requirements – human time and expertise for programming and documentation, machine time and storage requirements, overall time to completion and overall cost. In estimating the overall time it is essential to allow adequate time for program testing. The reliability of these estimates will, of course, depend on the experience of the programmer. If they are satisfactory then detailed 'internal' design of the program can begin. A flow-chart for the whole process is shown in Fig. 8.1.

A suitable general arrangement for a simple finite-element program is shown in Fig. 8.2. In the early stages of programming it is best to regard the three phases of input, calculation and output as quite separate, though later it may be more efficient to allow them to overlap. Program design,

Fig. 8.1. Some aspects of program design.

Limits on
(a) time and effort available for program development
(b) machine storage and program running time
(c) development and running costs
General computing experience

External program specification: numbers of nodes, elements, types of element, data and output formats
Theory of method and previous experience of programming it

Design thinking

User documentation
Program layout

Coding
Features of computer being used, programming languages available, existence of relevant subroutines

Modifications

User data
Testing ── Experience

Production runs

Results

Fig. 8.2. The arrangement of a simple finite-element program.

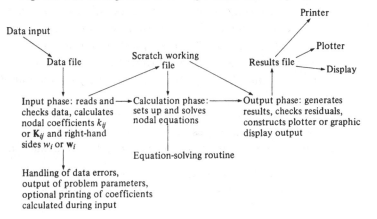

Printer

Data input
Plotter

Data file
Scratch working file
Results file
Display

Input phase: reads and checks data, calculates nodal coefficients k_{ij} or K_{ij} and right-hand sides w_i or $\mathbf{w}_i$
Calculation phase: sets up and solves nodal equations
Output phase: generates results, checks residuals, constructs plotter or graphic display output

Equation-solving routine

Handling of data errors, output of problem parameters, optional printing of coefficients calculated during input like design work in other fields, is an iterative activity in which both basic program structure and detailed coding are likely to alter as the work progresses. For example, decisions about program layout and storage requirements must be taken while there is still uncertainty about the coding required for the various phases and the amount of space it will occupy. These initial decisions may need revision as coding proceeds. The ability to get the overall structure of a program right at the first attempt comes only from experience.

A finite-element program is rarely an isolated static creation. A good programmer will try to work in such a way that new elements and other facilities can be added easily at a later date. All finite-element programs have a generic similarity and any program, once constructed, is potentially a useful starting point in the development of programs for other areas of application.

8.2   The input phase: computing the element properties

The input phase of a simple finite-element program† will typically consist of the following operations.

(a) Starting sequence. The program reads the title of the data, any global parameters such as the values of E and v and opens the scratch‡ and results files.

(b) Input of nodal coordinates. In its simplest form the associated section of the data consists of a list of node numbers followed by their coordinates in Cartesian or polar form. If the program allows higher-order elements to be specified it is useful to have a facility for interpolating nodes on the sides or faces of elements.

(c) Input of element descriptions. Again considering the simplest form, the associated data gives, for each element, the element type (triangle, quadrilateral, etc.), the number of nodes, the material properties (if these are different from the global values) and the global node numbers. The input program constructs the nodal stiffness matrices $\mathbf{K}_{ij}$ for each element and stores them on the scratch file. Since $\mathbf{K}_{ji} = \mathbf{K}_{ij}^t$ it is only necessary to compute those matrices for which $j \geqslant i$.

(d) Input of the external loads. The program computes the equivalent nodal loads $\mathbf{w}_i$ and stores them on the scratch file.

(e) Output of information abstracted from the data such as the

† To fix ideas a two-dimensional stress-analysis program is considered here.
‡ A scratch file is one used for temporary storage of intermediate results during the course of a calculation.

number of elements, number of nodal equations, etc. with optional output of quantities calculated during the input phase. The latter facility is useful if the program is to be used as an educational tool. It is also useful as a diagnostic facility during the development of the program.

It is desirable for any program to be able to detect errors and inconsistencies in the data supplied to it. Some of these errors can be detected as soon as the data item is read – an obvious example is an element defined as a quadrilateral but with only three node numbers listed. Other errors, such as gaps in the sequence of node numbers or the existence of disconnected elements, can only be detected after all the data has been read. In practice it is very difficult to anticipate all the types of error a program should guard against. The best way of developing a satisfactory data-checking system is to test the program on a variety of problems using data prepared by a group of inexperienced users.

When first considering the calculation of the nodal stiffness matrices $\mathbf{K}_{ij}$ it is natural to think in terms of a separate subroutine for each element type. However, the programmer who adopts this course of action will soon discover that much of the coding is the same for all types of element. The Gauss integration formula for the stiffness matrices of any two-dimensional element is

$$\mathbf{K}_{ij} = A' \sum_k h_k \mathbf{B}_i^t(\alpha_k, \beta_k) \, \mathbf{DB}_j(\alpha_k, \beta_k) \, |\mathbf{J}(\alpha_k, \beta_k)| \tag{8.1}$$

where α_k, β_k are the Gauss points and h_k are the Gauss weights. (For a constant-stress triangle only one Gauss point is needed.) The only quantities which vary between element types are

1. The range of element node numbers i and j.
2. The range of k and the values of α_k, β_k and h_k.
3. The shape functions n_i whose derivatives appear in $\mathbf{B}_i$ and $|\mathbf{J}|$.
4. The area A' of the generating element in the α, β plane.

It is appropriate, therefore, to construct a general routine, with these quantities as arguments, for the evaluation of all the nodal stiffness matrices. It is desirable to include in this routine a test on the computed value of $|\mathbf{J}|$. This can be used to detect elements with node numbers not in anti-clockwise order, re-entrant quadrilaterals and other element shapes likely to produce ill-conditioning. Note that at the heart of the routine is the evaluation of the repeated matrix product $\mathbf{B}_i^t \mathbf{DB}_j$. Although this calculation can be done by standard library subroutines it saves computing time if a special subroutine is written which takes advantage of the zeros in the matrices **B** and **D**, which have fixed positions in any particular problem type.

The generality of loading allowed is very much at the discretion of the writer of the program specification. In designing the program it is convenient in the first instance to restrict the input to concentrated nodal loads, any distributed loads being converted to equivalent nodal loads by hand. Routines to process more general load distributions can easily be added to the program at a later date. The equivalent nodal loads for any two-dimensional element carrying a distributed loading **w** are given by the Gauss integration formula

$$\mathbf{w}_i = A' \sum_k h_k \mathbf{w}(\alpha_k, \beta_k) \, n_i(\alpha_k, \beta_k) \, |\mathbf{J}(\alpha_k, \beta_k)| \qquad (8.2)$$

The amount of effort required to construct the parts of a finite-element program described in this section may be judged from the following figures, taken from an actual Fortran program for two-dimensional stress analysis.

Number of statements†

118 Main input program – reads title, opens files, reads nodal co-ordinates, element data and load data, checks data and prints information about problem.

119 Diagnostic program – prints error messages, gives optional printing of the element stiffness matrices and a 'map' of the nodal equations (see section 8.3).

66 General routine for the assembly of the matrices $\mathbf{B}_i$, **D** and **J** and the evaluation of the matrices $\mathbf{K}_{ij}$, in accordance with equation (8.1).

25 Subroutines for the evaluation of derivatives of the element shape functions:
three-node triangle (four statements),
four-node square (four statements),
six-node triangle (seven statements),
eight-node square (nine statements).

8 Subroutine for the evaluation of $\mathbf{B}_i' \mathbf{D} \mathbf{B}_j$.

50 General routine for the evaluation of the equivalent nodal loads $\mathbf{w}_i$ for a uniform distribution of load **w** throughout an element in accordance with equation (8.2).

30 General routine for the evaluation of the equivalent nodal loads $\mathbf{w}_i$ for a uniform distribution of load **w** on an element boundary. (These are given by an expression very similar to (8.2).)

† These numbers do not include DIMENSION, COMMON and similar 'background' statements.

25 Subroutines for the evaluation of the element shape functions:
three-node triangle (four statements),
four-node square (four statements),
six-node triangle (seven statements),
eight-node square (nine statements).

8.3   The assembly and solution of the nodal equations: band methods

The phase labelled 'calculation' in Fig. 8.2 involves the assembly and solution of the nodal equations to give the nodal displacements u_i. In the present section the assembly and the solution are considered as quite separate operations, carried out sequentially. In the next section procedures are described in which the operations are interleaved. In each section only methods based on Gaussian elimination are considered.

The central problem in programming Gaussian elimination is the efficient use of computer storage. The simplest approach is one in which the coefficients of the nodal equations are stored in a square two-dimensional array, with both dimensions equal to (or greater than) the number of nodal variables. This arrangement makes the programming of the solution routine very simple – indeed, the routine presented as the solution to problem 1.1 contains only 10 Fortran statements. It is also easy to construct the routine which transfers the element stiffness matrices K_{ij} and the nodal loads w_i from the scratch file to their appropriate places in the nodal equations, in accordance with (3.13b). However, except in trivial problems this procedure is very wasteful of computer storage, since all zeros in the equations are stored.

There are various ways in which the storage of zeros can be avoided. A convenient example for demonstrating some of the techniques is shown in Fig. 8.3a. The arrangement of the nodal equations for this example is shown in Fig. 8.3b. In this figure all non-zero submatrices† are marked with an X, while submatrices which are zero when the equations are first assembled but become non-zero as the calculation proceeds are marked with a 0. All other submatrices are initially zero and remain zero throughout the solution process.

If the equations are *symmetric* Gaussian elimination can always be organised in such a way that only coefficients on and above the leading diagonal are involved in the calculation. It follows that it is only necessary to provide storage for those submatrices which lie between the dotted lines in Fig. 8.3b. However, the programming is simplified considerably if the storage is 'filled out' to take the form of the trapezium indicated by heavy

† In a program for two-dimensional stress analysis these will be 2×2 submatrices.

lines in the figure. This gives each equation, apart from the last few in the 'toe' of the trapezium, the same amount of storage space. For a finite-element mesh with M free nodes the storage required is $(2M - B + 1)B/2$ submatrices, where B is the *bandwidth*, or maximum line length, given by

$$B = 1 + \frac{\text{Max. over}}{\text{all elements}} \left[\begin{array}{l} \text{Max. difference between global node} \\ \text{numbers associated with an element} \end{array} \right]$$

For the example shown in Fig. 8.3 the bandwidth is 7. Storage is required for 119 submatrices, in contrast to the 400 submatrices which make up the full set of coefficients. It is easy to evaluate the bandwidth during the input phase.

The 'housekeeping' statements associated with the above constant-

Fig. 8.3. Space requirements for constant and variable bandwidth solution schemes.
(a) The finite-element mesh.
(b) The arrangement of the nodal equations.

(a)

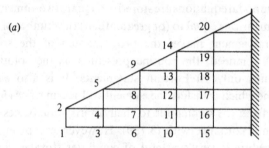

Bandwidth = 7

(b)

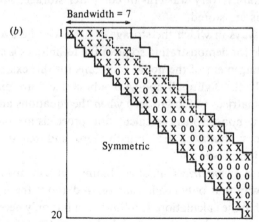

Symmetric

Storage
Full: 400
Constant bandwidth: 119
Variable bandwidth: 103

bandwidth storage scheme are rather more complex than those required with full two-dimensional array storage. However, the construction of an elimination routine using the scheme is well within the capacity of the average programmer – the equation solver in the program for plane stress analysis mentioned in the last section uses constant-bandwidth storage and contains only 35 Fortran statements. The use of banded storage also complicates the routine which assembles the equations. However, the difficulties are minor and in the plane-stress program this routine takes only 25 Fortran statements.

In writing the assembly and solution routines it is desirable to make the size of submatrices a variable parameter. This makes very little difference to the programming and allows the routines to be used unchanged in other finite-element programs, where the size of the submatrices may be different.

The efficiency of a constant-bandwidth equation solver is very dependent on the order in which the nodes of the finite-element mesh are numbered. Fig. 8.4 shows another numbering scheme for the mesh shown in Fig. 8.3 which gives a bandwidth of 10, with a corresponding storage requirement of 155 submatrices. A considerable amount of work has been done on the development of automatic node-numbering schemes which give minimum or near-minimum bandwidths.

Storage requirements may be reduced, at the cost of a considerable increase in program complexity, by changing to a *variable* bandwidth storage scheme. In such a scheme space is provided only for submatrices such as those indicated by an X or a 0 in Figs. 8.3*b* and 8.4*b*. The positions of these submatrices within the complete set of coefficients can be determined during the input phase. It is interesting to note that while the change from the node-numbering system of Fig. 8.3*a* to that of 8.4*a* increases the constant-bandwidth storage requirement by about 30%, the amount of storage required by a variable-bandwidth scheme is actually reduced slightly.

8.4   Interleaving the assembly and solution of the nodal equations: the frontal method

It is clear from Figs. 8.3 and 8.4 that when Gaussian elimination is applied to a set of nodal equations the elimination of one variable affects only a part of the complete coefficient matrix. Furthermore, the larger the number of nodes, the smaller the proportion of the coefficient matrix likely to be affected. It follows that it is possible to begin the elimination process *before* the complete set of equations has been assembled. Although this procedure does not reduce the total number of numerical operations required to solve a given set of equations it can lead to considerable savings in computer storage.

The idea can be developed in a number of different ways. The simplest has already been mentioned – the elimination of 'internal' nodes from elements such as nine-node quadrilaterals or ten-node triangles. This elimination can be carried out on an isolated element, since the only non-zero coefficients in the equations for an internal node are those associated with nodes on the boundary of that element. An obvious extension involves the assembly of several elements to form a 'super-element', followed by elimination of all the 'internal' nodes. The resulting

Fig. 8.4. The effect of a change in node numbering on the storage requirements.
(a) The finite-element mesh.
(b) The arrangement of the nodal equations.

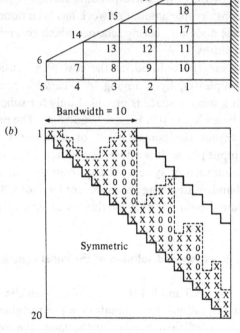

Storage
Full: 400
Constant bandwidth: 155
Variable bandwidth: 95

element can then be connected to other elements or super-elements in the usual way. This procedure is simply the finite-element version of the 'substructure' concept, which was introduced into skeletal structural analysis in the early 1950's (see reference 2, chapter 11).

The most sophisticated development of the idea is the *frontal* method of solving a set of nodal equations. This is essentially a version of Gaussian elimination in which the coefficients in the equations are assembled and processed in a sequence determined by the advance of a wave-front through the finite-element mesh. In contrast to the methods described in section 8.3 the sequence followed by the frontal method is determined by the order in which the *elements* are specified, rather than by the order in which the *nodes* are numbered. The advantage of the method lies in the fact that at no time during the calculation is the complete coefficient matrix present in the computer store. For the example shown in Fig. 8.5 the procedure is as follows.

The processing of the elements is assumed to take place in the order A, B, C, D.... The first step is the addition of the coefficients for element A to the space allocated to the nodal equations, in accordance with equation 3.13b. Since node 1 is not connected to nodes in other elements the two scalar equations for node 1 are complete, and since the equations are symmetric the coefficients in the equations for node 1 give the row-multipliers required for the elimination of the displacements associated with node 1. Thus the correct multiples of the equations for node 1 may be added to the equations for nodes 2, 3 and 4, even though the latter equations are still incomplete. Once the additions have taken place the equations for node 1 play no further part in the elimination and may be transferred to subsidiary storage ready for the final back-substitution phase of the solution process. At this stage, the 'wave-front' follows the boundary 2–4–3, as shown in Fig. 8.6a. Notes on the wave-front are termed *frontal nodes*.

Fig. 8.5. Element ordering for a frontal solution scheme.

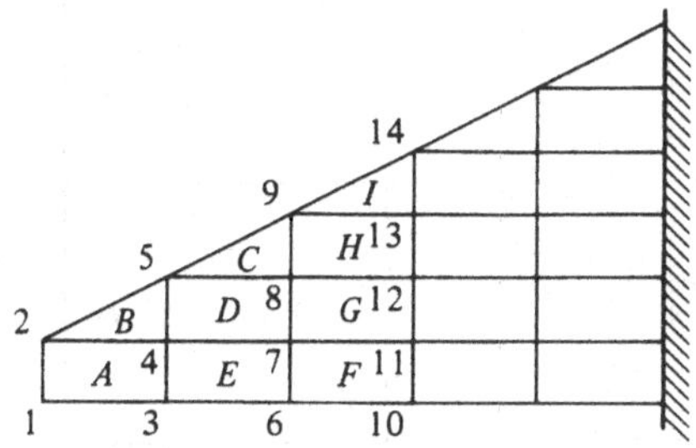

The addition of element *B* completes the equations for node 2, so that the displacements associated with this node can be eliminated and the equations transferred to subsidiary storage. Node 5 is added to the wave-front as shown in Fig. 8.6*b*.

The addition of element *C* causes coefficients to be inserted in the equations for nodes 8 and 9, but does not complete the equations for any of the active nodes. No elimination takes place, but nodes 8 and 9 are added to the wave-front, as shown in Fig. 8.6*c*. The addition of element *D* adds node 7 to the wave-front and allows the elimination of the displacements associated with node 5, and so on.

When all the elements have been processed the back-substitution phase of the solution scans through the nodes in an order which is the reverse of the order in which they were eliminated.

From this example it is apparent that the programming of the frontal method involves considerably more 'housekeeping' than the band methods described in section 8.3. A Fortran program is likely to involve at least

Fig. 8.6. The elimination sequence in a frontal solution scheme.

	Element introduced	Nodes eliminated	Frontal nodes
(a)	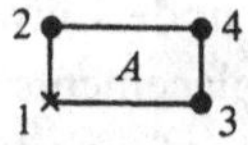	1	2 − 4 − 3
(b)	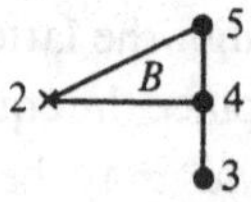	2	5 − 4 − 3
(c)	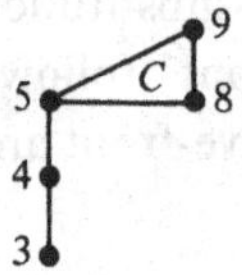	none	9 − 8 − 5 − 4 − 3
(d)	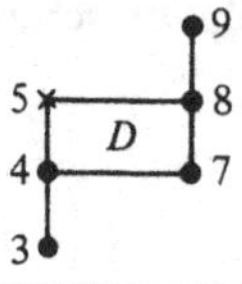	5	9 − 8 − 7 − 4 − 3
(e)	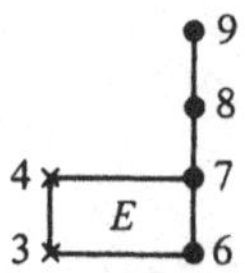	3, 4	9 − 8 − 7 − 6

500 statements and its construction is not a task to be undertaken by inexperienced programmers. However, its advantages make it the method most commonly used in professionally-written finite-element programs. In the frontal method the 'front-width' (i.e. the maximum number of frontal nodes) depends on the order in which the elements are scanned. Automatic procedures for front-width minimisation have been developed which are similar in effect to the schemes for bandwidth minimisation mentioned in section 8.3.

8.5   The output phase: program testing

By the time a start is made on the coding of the output phase most of the difficult programming will have been done. Indeed, the essential output associated with a finite-element stress-analysis program – the printing of displacements and stresses, requires very little new coding. The nodal displacements u_i are provided by the calculation phase, while routines for evaluating the element shape functions n_i are available from the input phase. It is therefore straightforward to evaluate the displacement $u = u_i n_i$ at any point in the solution region. The stresses $\sigma = DB_i u_i$ are also easily evaluated, since coding for the construction of the matrices D and B_i is also available from the input phase. As mentioned in section 4.8, it is common practice to evaluate the stresses at the Gauss points rather than at the nodes (where stress discontinuities are likely to occur).

An equilibrium check is often included in the output phase. This involves comparing the equivalent nodal forces $K_{ij} u_j$ derived from the computed nodal displacements u_j with the equivalent nodal forces w_i calculated from the loading data. The matrices K_{ij} are, of course, already available from the input phase.

In the early days of computing an equilibrium check was an important safeguard against machine hardware errors. Such errors are now rare and are in any case detected by most computer operating systems. In a present-day finite-element computer program the function of an equilibrium check is merely to give the user information about the overall accuracy of a solution. The out-of-balance nodal forces should be of round-off order in comparison with the applied loads. Any excessive out-of-balance indicates that the nodal equations are ill-conditioned. The cause is likely to be an injudicious mesh arrangement or, of course, an error in the input data.

It is important to realise that an equilibrium check only tests the accuracy of the solution of the nodal equations. Since the K_{ij} matrices are used both in setting up the equations and in applying the check, an error in one of these matrices will not affect the out-of-balance forces unless it is such as to make the equations ill-conditioned. Nor can an equilibrium check indicate the accuracy of the finite-element modelling process itself, or detect data errors which result in the analysis of an incorrect (but physically possible) problem. In the program for two-dimensional stress analysis described earlier the printing of displacements and stresses required 30 Fortran statements (in addition to those taken from the input phase) while the equilibrium check required 40 additional statements.

The remainder of the output phase is essentially a matter of presenting the results in convenient form. It may include a variety of graphical output – a mesh plot, a displacement plot, principal stress contours, etc. A plot of the finite-element mesh is particularly useful as a check on the geometrical correctness of the input data. The effort required to code these operations is very dependent on the amount of general graphics software available to the programmer.

It is one of the basic principles of good programming that all individual routines should be thoroughly tested before being connected together. However, tests must also be run on the complete program. The following suggestions may be useful to readers planning the testing of a new finite-element program for elastic stress-analysis.

1. If possible, include a pin-ended bar in the list of available elements. The evaluation of the relevant stiffness matrices is very simple to code and allows testing of most of the organisational parts of the program. In contrast with continuum analysis the stressing of a pin-jointed frame involves no discretisation approximations, and many examples of solutions are available in textbooks.

2. Check that a solution produced by the program is independent of the orientation of the coordinate axes, the numbering of the nodes and the sequence in which the elements are specified.

3. Check that the program produces symmetric results when presented with data describing a symmetric system.

4. Check that the program produces the exact solution under conditions of constant stress (assuming that all the elements available should, theoretically, satisfy this condition).

5. Check results against published results produced by other finite-element programs. Note, however, that there will inevitably be small discrepancies due to changes in computer word-length and numerical procedures. It may be difficult to decide whether such

discrepancies are large enough to indicate an actual programming error.

6. Check that the program runs satisfactorily on the maximum size of problem specified in the program-user's guide. Mistakes in storage allocation may not show up until a problem reaches a certain size.

Finally it must be emphasised that a user should never have unquestioning trust in the results of any finite-element analysis. Results should always be assessed in the light of common-sense and general engineering judgement, and rough manual analyses should be carried out for comparison wherever possible.

8.6 Commercial finite-element programs

In the earlier sections of this chapter it has been argued that the construction of a finite-element program is well within the capacity of anyone with a reasonable amount of computing experience. A program which can carry out the essential steps of a finite-element analysis, as described in the earlier chapters of this book, need consist of no more than a few hundred statements in a high-level language – quite a small amount of coding by professional standards. However, there is a considerable difference between a program written for private use and one written as a production tool serving a wide spectrum of users. The present section describes some of the additional features which one may expect to find in a commercial finite-element program.

Most such programs are 'general-purpose' in the sense that they provide facilities for solving a number of different kinds of physical problem. This is not simply to make efficient use of those routines, such as the one for solving the nodal equations, which are independent of the problem type. It also allows a user to carry out analyses where two or more kinds of field interact, such as the calculation of stresses due to thermal strains or dynamic fluid forces. In such cases it is clearly advantageous if the results of one part of a calculation (the computed temperature field, for example) can be handed over as input data to another section of the program without any manual transcription.

The account of the finite-element method given in this book is restricted to linear problems. Many users of the method require non-linear analyses, and appropriate facilities are provided in most of the larger programs. These cover problems involving non-linear materials (plasticity, fracture, ferromagnetism) and gross changes of geometry (extrusion, buckling, moving boundaries). The iterative approach used in such analyses can also be applied to dynamic problems and those involving history-dependent materials. In all these problems the solution of the nodal equations has to be repeated after each increment of time or loading. It is important, therefore, that the solution procedure is coded to be as efficient as possible.

When a finite-element program is being designed it is natural to make the limits on problem size as large as possible. Although the proportion of problems which really need several thousand nodes and elements is small, it is these problems which receive publicity and establish the reputation of a program. Programs which cater for 'large' problems almost invariably use a frontal equation-solver, with facilities for the construction of sub-structures and super-elements, as mentioned in section 8.4. Note that the computing problems associated with large systems of equations are not only those concerned with efficient use of storage. Cumulative round-off errors also build up as the number of nodes increases and it is likely that in a large problem certain parts of the calculation will require double-precision arithmetic.

Several commercial programs provide facilities for analysing structures which consist of a number of identical sub-structures arranged in a symmetrical fashion – a multi-pole electrical generator and a domed roof with radial ribs are typical examples. Fourier techniques similar to those described in chapter 6 allow the analysis of a complete symmetric structure to be replaced by a series of analyses of a single sub-structure, thus reducing the problem size, though not necessarily the computing time.

The following details, which relate to the PAFEC system, give some idea of the capacity of a large commercial finite-element program and the effort involved in its construction.

PAFEC can carry out stress analyses under static, transient and free-vibration conditions. The materials involved in the analysis may be linear-elastic or may obey more complex constitutive laws, including those associated with anisotropy, creep and plasticity. It is possible to include thermal strains and large-deformation effects (including buckling), and the Fourier decomposition techniques described in chapter 6 may be used in problems involving cyclic symmetry.

Over 80 types of finite element are available, including virtually all those mentioned in this book. Groups of elements may be combined to form sub-structures, which can then be treated simply as additional types of element: several levels of sub-structuring are available. The data-input section of the program applies extensive checks to the user's data. Both banded-matrix and frontal solution procedures are available, with automatic element re-sequencing to give maximum efficiency in the frontal solution.

No specific limits can be stated for the size of problem which can be solved by a program such as PAFEC, since the limits depend on the particular computer on which the program is run. The complete PAFEC suite of programs consists of approximately 250000 Fortran statements. Roughly 17% of these are associated with the element library, while 31% are concerned with assembling and solving the nodal equations. The remainder cover such operations as data checking, sub-structuring, non-linear and cyclically symmetric analyses and graphical output. The program represents approximately 200 man-years of effort.

It is a mistake to imagine that any large program can be written and then just handed over to its users. A system the size of PAFEC is always being extended, and technical support must be provided for users of the program on a large number of different types of computer. This support must include the supply of up-to-date documentation as well as the maintenance of the program itself. The development and support of PAFEC currently involves about 20 full-time staff.

As finite-element programs such as PAFEC have developed they have acquired auxiliary programs known as *pre-processors* and *post-processors*. A pre-processor is a program designed to simplify the preparation of data. The simple input scheme described in section 8.2 requires the specification of each individual nodal coordinate and element boundary. Anyone who has carried out this task, even for a simple two-dimensional problem with a few tens of nodes, will know how tedious and prone to error the process is. For a three-dimensional problem with a complex boundary the preparation and checking of the data becomes a serious practical problem.

The obvious solution to this difficulty is the development of a program which will generate the finite-element mesh. Such a program can be organised in two alternative ways. In *interactive* mesh generation the user sketches the mesh on a graphics terminal with a light-pen or similar device and the program picks off the positions of the nodal coordinates and the element boundaries. In *automatic* mesh generation the user simply defines the boundary of the solution region, the number of nodes or elements required and the types of elements to be used. The program then fills the solution region with the number of elements specified, avoiding if possible the introduction of grossly distorted elements. The best procedure is probably a combination of the two schemes, with the user defining a coarse mesh interactively and the program sub-dividing the mesh automatically when required. Any form of automatic mesh generation requires some

Fig. 8.7. The generation of a three-dimensional finite-element mesh
(i) Data defining the geometry of the complete object is used to
produce a perspective view with hidden lines removed.
(ii) Symmetry allows the analysis to be carried out using a 60° sector.
A modified version of the initial data is created and used to produce a
perspective view of the sector.
(iii) A coarse finite-element mesh for the sector is generated and
plotted using the stored geometrical data.

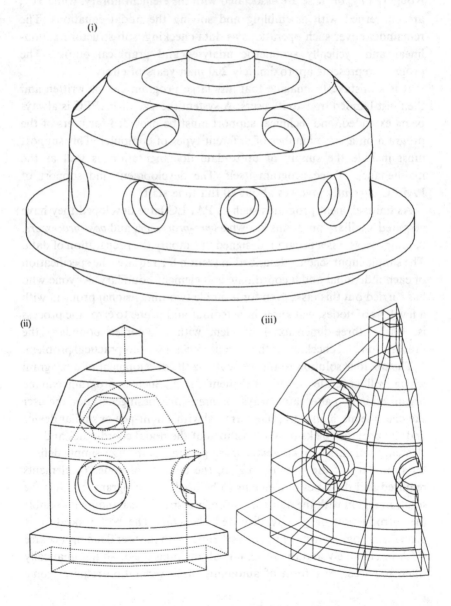

(i)

(ii)

(iii)

(iv) The elements are subdivided to give a finer mesh. In this example the complete mesh formed from the six sectors has 1586 elements and 10132 nodes.

(v) Auxiliary surface (*a*) and section (*b*) views are produced to aid visualisation of the full three-dimensional mesh shown in (iv). The program user has facilities for making manual alterations to the machine-generated mesh if the latter contains any grossly distorted elements.

(iv)

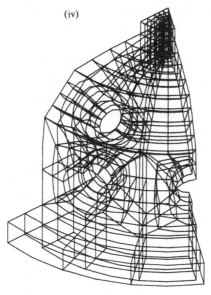

(v)(*a*)

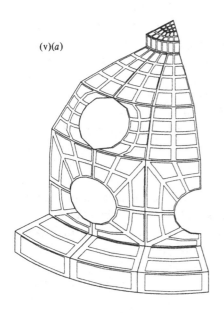

(v)(*b*)

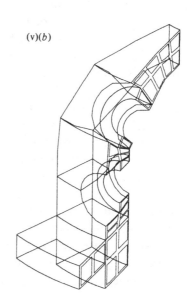

scheme of automatic node numbering or element sequencing – if possible a scheme which minimises the time required for solving the nodal equations.

A simple and flexible mesh-generation scheme makes it possible to carry out a rapid preliminary finite-element analysis of a problem using a coarse mesh. This analysis, although not accurate, indicates those parts of the solution region where there are rapid variations in stress, temperature, etc. The mesh can then be refined in those regions to give a more accurate analysis.

A satisfactory solution to the mesh-generation problem uncovers the next problem associated with data input – the best way of specifying the boundaries of the solution region. Most finite-element analyses are concerned with man-made objects – ships, nuclear reactors, machine tools, etc. The shapes of these are built up from simple geometrical forms – planes, cylinders and spheres in the great majority of cases, and it is dimensions associated with these forms which specify the object, both to its designer and to those who will eventually manufacture it. There are considerable advantages if a finite-element program can accept the same specification.

A short-term advantage is that the input data defining the object to be analysed can be taken directly from dimensions on the designer's drawing, thus reducing the volume of data and minimising the risk of data errors. Much more important, however, is the opportunity which is created for linking the finite-element analysis to the design process. If the data supplied to the computer program consists of a specification of the object in the normal language of the designer, he can check on the effect of changing a particular dimension simply by altering a single number in the data and repeating the analysis. A sequence of plots showing the generation of a three-dimensional finite-element mesh from a geometrical description of an object stored within a computer is shown in Fig. 8.7. These plots were produced by the ROMULUS and FEMGEN programs.

An efficient and versatile pre-processor makes a finite-element program very much easier to use. However, satisfactory integration of a finite-element analysis into a design procedure also requires a *post-processor* to convert the voluminous numerical output from the analysis into a suitable basis for design decisions. This program may well correlate the results of a series of analyses, identifying the most serious load cases and indicating positions of maximum stress, etc. A post-processor, like a mesh-generator, is likely to make considerable use of computer graphics.

The final picture, then, is of a finite-element program as just one part of a computer-aided design system, the whole system being controlled interactively by the designer from a terminal – preferably one with graphical display facilities. Such a system holds a single description of the object being designed, in a form suitable for processing by the various sub-systems – finite-element program, graphical display, flat-bed plotter for producing manufacturing drawings, program for generating tape output to a numerically-controlled machine tool, etc. If the finite-element mesh is generated automatically and is erased before contours of stress, temperature, etc. are plotted, the designer may hardly be conscious that the results he sees have been generated by a finite-element analysis. In a situation where layers of sophisticated computer programming separate the designer from the mathematical details it becomes ever more important for critical engineering judgement to be applied to the final output from the system.

REFERENCES

The following texts provide further information about topics dealt with in this book.

1. *The finite element method*, O. C. Zienkiewicz, 3rd edn., McGraw-Hill, 1977.
   This is one of the best-known texts on finite-element analysis. It covers virtually all the material presented in the present book and much else besides.
2. *Matrix methods of structural analysis*, R. K. Livesley, 2nd edn., Pergamon, 1974.
   This is mainly concerned with the analysis of skeletal structures, i.e. frames, trusses and arches.
3. *Vector fields*, J. A. Shercliff, CUP, 1977.
   This gives a good introduction to the concepts of vector calculus, with many examples of its application to problems in engineering and physics.
4. *Theory of elasticity*, S. P. Timoshenko and J. N. Goodier, 3rd edn., McGraw-Hill, 1970.
   This gives background material on the more traditional aspects of stress analysis, with emphasis on analytical rather than numerical solutions.

NOTATION

∇	gradient operator
$\nabla\cdot$, ∇^t	divergence operator
u	scalar field
$\mathbf{v}$	vector field
$\mathbf{a}, \mathbf{b}$	vectors, components $a_x, a_y, ..., b_z$
t	transpose superscript
$i...m$	dummy subscripts (italic)
p...s	specified subscripts (roman)
$u(x)$	function of x
M	number of integration points or functions
$c_0, ..., c_{M-1}$	set of M constants
$u_1, ..., u_M$	function values
n_i	interpolating polynomials/shape functions
δ_{ij}	Kronecker delta
P	dimensions of Euclidean space
$f(x)$	arbitrary function of x
α, β	parametric coordinates
$\mathbf{J}$	Jacobian matrix
R, R'	regions of integration
h_i	Gauss weights
$\mathscr{L}$	differential operator
$w(x)$	known right-hand side of differential equation
$\phi_i(x)$	Ritz functions
λ	eigenvalue
$\tilde{u}(x)$	true solution
H	horizontal force
$U(x)$	set of functions $u(x)$
T	functional

W	loading parameter
k_{ij}	array of nodal coefficients
Δu	variation in u
ϕ_s	smoothed function
ε	small interval
EI	flexural stiffness
c	constant
F	vertical force
C	boundary value of u
a, b	range of x
$\varepsilon(x)$	error function
γ_i	set of weighting functions
$\delta(x_i, x)$	delta function
h	interval in finite difference mesh
i	current
v	voltage
r	resistance
g	conductance
I	known nodal currents
$\mathbf{w}$	loads
$\mathbf{d}$	displacements
$\mathbf{K}_{ij}$	stiffness matrices
L	length
ρ	density
A	cross-sectional area
g	gravity acceleration
u	temperature or electrical potential
$\mathbf{e}$	gradient of temperature or potential
$\mathbf{q}$	density of heat or current flow
D	conductivity
w	source distribution
M	number of nodes
R	solution region
$\mathbf{r}$	position vector
n_j	element shape functions
A	area of element
L_j	half-lengths of sides of triangular element
$\mathbf{n}_j$	unit normals to sides of triangle
$\mathbf{b}_j$	gradients of shape functions
q_j	equivalent nodal inflows
k_{ij}	nodal coefficients
w_i	equivalent nodal sources
A_i	areas associated with nodes

u_i	nodal values of u
A_p	area associated with node p
N_m	nodal shape functions
S	boundary of solution region
i', j'	global node numbers associated with nodes i, j
γ_m	set of independent functions
S_p	boundary of area round node p
$\bar{N}_m$	'top-hat' nodal shape functions
s	*italic* dummy subscript for set of boundary nodes
V	volume of three-dimensional element
$\mathbf{D}$	conductivity tensor
E_z	component of electric field
c	specific heat
$\mathbf{u}$	displacement field
u_x, u_y	components of displacement
ε	strain field
$\varepsilon_{xx}\ldots$	direct strain
$\gamma_{xy}\ldots$	shear strain
$\square$	matrix strain operator
$\mathbf{D}$	material properties matrix
E	Young's modulus
ν	Poisson's ratio (nu)
σ	stress field
$\sigma_{xx}\ldots$	direct stress
$\tau_{xy}\ldots$	shear stress
$\mathbf{w}$	load distribution
w_x, w_y	components of load
S_1	segment of boundary
$\mathbf{u}_m$	nodal displacements
$\mathbf{u}^*$	virtual displacement field
ε^*	virtual strain field
$\mathbf{B}_i$	nodal matrices
$\mathbf{K}_{ij}$	nodal stiffness matrices
$\mathbf{w}_i$	equivalent nodal loads
$\tilde{u}_r$	specified (boundary) nodal displacement
$\mathbf{n}$	unit normal to boundary
$\mathbf{t}$	unit tangent to boundary
K	axial stiffness
α	angle
$\tilde{u}$	true displacement field
r	distance from point load
θ	temperature
α	coefficient of thermal expansion

ε_θ	temperature strain field
$\mathbf{M}_{ij}$	mass matrices
p	mass per unit area
P	degree of polynomial
x', y', z'	new Cartesian coordinate system
α, β, γ	local coordinates
Δ	specified displacement component
$\mathbf{u}^L$	linear displacement field
$\mathbf{C}_1$	constant matrix
Ω	small rotation
I	moment of inertia
b, d	dimensions of cross-section
h	element size
n_i^L	set of linear shape functions
$\bar{\mathbf{q}}$	mean flow
r, ϕ, z	cylindrical polar coordinates
$\bar{r}$	radius of centroid
r_0	radius to origin of local coordinates
n_i^M	set of mapping functions
(k)	superscript indicating number of harmonic
$\bar{\mathbf{b}}_i$	modified vector
$\bar{\mathbf{B}}_i$	modified matrix
s, c	subscripts indicating sine and cosine components
ω	angular velocity
θ	rotation of normal
m	moment
b, s	subscripts denoting 'bending' and 'shearing' terms
G	shear modulus
κ	curvature (kappa)
$\mathbf{T}$	transformation matrix
α	angle of rotation
$\boldsymbol{\theta}$	rotation vector
t	plate thickness
$\mathbf{m}$	moment vector
$\boldsymbol{\kappa}$	curvature vector (kappa)
$n_{xx}, n_{\phi\phi}$	tangential force components
M	number of free nodes
B	bandwidth

INDEX

[A reference in italics refers to a question at the end of a chapter, the chapter number being placed first.]